D1524341

Economics and Ecology

United for a Sustainable World

Social-Environmental Sustainability Series

Series Editor

Chris Maser

Published Titles

Economics and Ecology: United for a Sustainable World,
Charles R. Beaton and Chris Maser

Sustainable Development: Principles, Frameworks, and Case Studies,
Okechukwu Ukaga, Chris Maser, and Michael Reichenbach

Social-Environmental Planning: The Design Interface Between Everyforest and Everycity, **Chris Maser**

Economics and Ecology

United for a Sustainable World

Russ Beaton and Chris Maser

CRC Press is an imprint of the
Taylor & Francis Group, an **informa** business

CRC Press
Taylor & Francis Group
6000 Broken Sound Parkway NW, Suite 300
Boca Raton, FL 33487-2742

Printed in the United States of America on acid-free paper
Version Date: 20110608

International Standard Book Number: 978-1-4398-5295-8 (Hardback)

Visit the Taylor & Francis Web site at
http://www.taylorandfrancis.com

and the CRC Press Web site at
http://www.crcpress.com

Dedication

I dedicate this work to my grandchildren, Alia Woodyard, Seth Beaton, Jaxson Woodyard, and Caden Beaton, and to all other grandkids, now and in the future. May they have the same opportunity we have had to enjoy the bounty of this marvelous planet.

Russ (RB)

I respectfully dedicate this book to the memory of my father, Clifford Elgis Maser (1910–1968), Dean of the School of Business and Technology at Oregon State College/Oregon State University from 1947 to 1966.

Chris (CM)

Contents

Section III Reconciliation and Looking to the Future

Series Editor Note

All systems, those of nature and those of human construct, are governed by nature's biophysical principles. In turn, the two laws of thermodynamics and the law of maximum entropy control the biophysical principles. These laws and principles are inviolate, yet people who lack a sense of inner security—and thus fear life—can never have enough monetary wealth to feel safe from the specter of uncertainty. Thus, instead of living within the constraints of nature's laws and biophysical principles through cooperation and sharing, they attempt to subvert these laws and principles to their personal benefit through economic competition and the greed that spawns it. When, however, subversion does not work, they simply ignore nature and discount the welfare of all future generations as irrelevant. Clearly, an economic system is severely broken that allows such blatant disregard for its operating principles of mutual survival and well-being.

Alas, merely throwing money at our broken economic system in a bid to treat symptoms and continue with business as usual will not work. It never has. But, to mend what is broken, we must reach beyond where we feel safe and dare to move ahead, despite the fact that perfect knowledge will always elude us. There are no biological shortcuts, technological quick fixes, or political hype embodied in our current symptomatic thinking that can mend what is broken. Dramatic, fundamental change in the form of *systemic thinking* is necessary if we are to have an economic system able to serve as an instrument of social-environmental sustainability based on the triple bottom-line of *ecological integrity, social equity,* and *economic stability.*

Chris Maser
Series Editor

Preface

Earth is our home, and it is in crisis. There are two sides to this crisis—our economy on the one hand, and its effect on the ecology of our home planet on the other. Our economy, the human-constructed part of our home, represents the collection of manufactured artifacts, institutions, and technologies with which the human species, over the millennia, has designed with pride and used with arrogance to wrest a material existence from a wondrously bountiful but ultimately finite planet.

Despite conventional thinking in the industrialized world that typical monetary and fiscal manipulations will restore order and put us back on the path of economic growth, our economy is in danger of collapse. Furthermore, over the last few decades, we have experienced incessant globalization of the resource, manufacturing, and financial systems; and this serves to extend the threatened collapse to the entire industrialized world, and not simply to the economy of the United States.

Meanwhile, the natural environment is similarly sending unmistakable warnings. Glaciers melt, oceans are becoming dangerously acidic, species become extinct, natural habitats are diminishing, and weather patterns become increasingly severe and unpredictable each year. The stress on resource systems of all kinds threatens to shrink the carrying capacity of the planet, even as we call upon it for increased contributions to support a burgeoning human population.

The approach of those in power, whether their base is in the public or private sector, has historically been to particularize such problems, isolate and treat each as separate issues, and attempt band-aid measures acceptable to a public oriented to business as usual. This attitude is derived through the linear, mechanical thinking that addresses symptoms rather than the systemic causes, which means the ultimate goal of policy is to restore conventional economic growth. Even in the best of times, before problems reach crisis proportions, this economic drive ignores both the seriousness and the interconnectedness of issues that might arise.

We are convinced that the problems must be treated systemically if they are to be effectively treated at all. The academic fields in which we were trained (Beaton as an economist and Maser as an ecologist) each represent a vital component of the world crisis that has developed by placing the "natural" in opposition to the "produced." We cannot, in good conscience, fail to speak out at this time. It is our fervent hope that our collaboration, our third, grounded in a deep respect for the perspective and contribution of the other, can shed useful light on the problems we—and all generations to come—now face.

It is our observation that the necessary message could not be more urgent. Indeed, the only preferable timing would be that it could have been earlier.

The problems worsen continuously, and even the conventional solutions, though they may appear to offer temporary relief, will in many ways make the long-run day of reckoning even more painful.

In the strictest sense, this is not a book about policy. We propose no new laws, regulations, or agencies to deal with the issues. We recommend no sweeping international agreements or conferences. Instead, we suggest attitudes and characteristics that might someday infuse our laws and institutions in an ideal world. But the world cannot wait for such top-down reform, as slow as it always is. Rather, this is a book that speaks to the individual. Effective first steps to a sustainable world must begin with enlightened citizens, and the real need is for a different systemic worldview in support of the bottom-up change that is, in our opinion, most lasting and effective.

We are moved by the words of Margaret Mead: "Do not doubt for a minute that a small group of committed people can change the world. Indeed, that is all that ever has."* Perhaps then, in the final reckoning, this book *will* be about policy—a policy that begins in the hearts and minds of regular, everyday citizens living in their communities and including in their actions and deliberations the voices of their children and the generations of children yet unborn. That is for whom we speak.

* Brainy Quotes, Margaret Mead, http://www.brainyquotes/authors/m/margaret_mead.html.

Acknowledgments

Any book that is coauthored must first acknowledge one's coauthor. To my good friend and colleague, Chris Maser, I owe the inspiration for this entire project. Earlier in life, neither of us thought we would chance to meet someone with both the professional background and the personal point of view that we have found represented in each other. This collaboration is a labor of love and deep friendship. No idea is off limits, and each new insight seems to inspire creativity in the other.

Next, nothing would have happened without the encouragement and understanding of my wife, Delana. She is a ready and willing partner, both with ideas and editing, and is always understanding of the constraints involved in taking on such a project. I promise that I will have more time this year for the garden and the grandchildren.

Our daughter, Lynn, has served as both the environmental policy advisor and as deputy director of the Oregon Economic and Community Development Department, where she was instrumental in authoring a Sustainability Executive Order for each of the last two governors of Oregon. She has always offered a very informed and helpful window into the real world of public policy concerning these issues. Further, it was through her that Chris Maser and I met some 15 years ago, and we are both forever grateful.

This book is a product of much reading, teaching, and policy involvement over the years. Thus, I owe much to my students at Willamette University (Salem, Oregon)—too numerous to mention—who always willingly provided contributions to the ongoing debates of our day. If they learned as much from me as I did from them, it will have been a successful exchange.

Professionally, my role models include the likes of E. F. Schumacher, Herman Daly, Hazel Henderson, David Korten, Paul Hawken, and Amory Lovins. I have had the pleasure of meeting and interacting with each of them at one time or another. Their words and their writings offer inspiration and clarity to anyone seeking to create a better future for humankind. No one could read even a sampling of their collected writings and come away unchanged.

Finally, we both owe a debt of gratitude to Ellen, the friendly waitress at the Pine Cone Café, just off Interstate 5 near Albany, Oregon (between our homes in Salem and Corvallis), where Maser and I often met for breakfast and engaged in long and animated discussions as this book was being written. We hope we did not disturb your other customers (mostly truckers) and that you did not mind us occupying the table by the window long after the breakfast was eaten. And the eggs were always just right.

RB

I thank Russ Beaton for the privilege of working with him again. I am also much indebted to my beautiful wife, Zane, for her continuing patience as I worked on this book.

CM

Authors

Russ Beaton received his bachelor's degree from Willamette University in Salem, Oregon, and his Master's and Ph.D. degrees from Claremont University, California. His original training was in mathematical economics and econometrics, although his doctoral thesis was in location theory and urban land economics, which became a lifetime interest—one of his many interests in which he has no formal training.

After teaching for 3 years at California State College at Fullerton (now Fullerton State University), and 4 years at Simon Fraser University in Vancouver, British Columbia, Canada, Beaton returned to his alma mater, Willamette University, where he taught economics and did research for 33 years. He has also taught in four graduate business (MBA) programs over the years, helped design two of them, and served as director of one.

His teaching interests included Microeconomic Theory, Environmental Economics, Energy Economics, History of Economic Thought, and Regional Economics and the Economy of Oregon—all courses he developed and introduced, and which contributed directly to the issues covered in this book.

Beaton's research interests have always gravitated toward useful policy efforts that have the capability of directly and immediately affecting the lives of people, even though early training was in the purely theoretical realm. He has consulted and done policy-based contract research for at least six different agencies of the State of Oregon, in areas such as land use, agriculture, timber, transportation, energy, housing, and general economic policy.

Beaton is coauthor with Chris Maser of two other books and participated in drafting the legislation, passed by the 1973 Oregon Legislature, that created Oregon's widely acclaimed land-use planning system.

He and Delana, his wife of 49 years, live on a small farm outside Salem, Oregon, where they care for a few animals, garden organically, enjoy their three children and four grandchildren, harvest solar energy, and attempt to learn and demonstrate to others as much as possible about the desirable characteristics for a sustainable future.

Chris Maser was trained in zoology and ecology and worked for 25 years as a research scientist in agricultural, coastal, desert, forest, valley grassland, shrub steppe, and subarctic settings in various parts of the world before realizing that science is not designed to answer the vast majority of questions society is asking it to address. Science deals with an understanding of biophysical relationships; society asks questions about human values.

Maser gave up active scientific research in 1987 and has since worked to unify scientific knowledge with social values in helping to create sustainable communities and landscapes, part of which entails his facilitating the resolution of social-environmental conflicts. He has contributed to more than 286 publications, including 34 books, mostly dealing with some aspect of social-environmental sustainability. In addition, he is listed in *Who's Who in the West, American Men and Women of Science, Contemporary Authors,* and *International Authors and Writers Who's Who.*

As an author, international lecturer, facilitator in resolving environmental disputes and creating vision statements, and assisting in sustainable community development, Maser is committed to telling it as he sees it in order to give people an honest, open appraisal, based on extensive experience, of innovative ways to resolve their conflicts and unite their community in a sustainable way in relationship to their surrounding landscape whence comes the community's natural wealth. This is important because people can most easily deal with required changes when they feel safe and know someone is there to guide the process as gently as possible, all the while protecting their dignity and looking out for their well-being.

Although he has worked and lectured in Canada, Egypt, France, Germany, Japan, Malaysia, Mexico, Nepal, Slovakia, and Switzerland, he calls Corvallis, Oregon, home.

Introduction

[Americans'] one primary and predominant object is to cultivate and settle these prairies, forests, and vast wastelands. The striking and peculiar characteristic of American society is, that it is not so much a democracy as a huge commercial company for the discovery, cultivation, and capitalization of its enormous territory.... The United States is primarily a commercial society... and only secondarily a nation.[1]

French political scientist Émile Boutmy
(April 13, 1835 to January 25, 1906)

Social and Natural Systems in Simultaneous Crisis

The natural world is in revolt. In the everyday bustle of life in our consumer-oriented culture, it is easy for harried citizens to be unaware of this disturbing fact. They do not see the tidewater-glaciers receding in semipolar regions of the world. They are not directly aware of the dramatic paths of extinction on which a multitude of plant and animal species are already bound. They do not see the desecration of tropical rain forests and the resultant effects on weather patterns. They and their families have not personally experienced the impact of new, exotic diseases whose spread augments the impacts of expanding climate change. Even when evidence of the tremendous stress our home planet is under is clearly presented, the roots of informed denial run deep: "I haven't seen it personally." "There is conflicting evidence." "There is no conclusive, scientific proof." "These trends are global. I have no power to change them." "If it's true, 'they' will do something about it." But, the eternal question remains: Who is "they?"

The media, along with our political system, has not helped. Most environmental news, including that of climate change, is presented in the most sensationalistic manner possible, one guaranteed to generate controversy. Stimulating an audience for the particular media outlet seems much more important than ensuring accuracy, avoiding bias, or informing citizens of true and present dangers. Politicians have willingly bought into, if not helped create, the assumption that "our economy versus the environment" is the appropriate conflict. Controversy has been created, to generate readership or votes, by focusing on issues where little real controversy would exist if all the evidence were thoughtfully considered. Paid experts are supported to promulgate public skepticism and thus protect some status quo, business-as-usual activities.

As a result, the public is besieged with a welter of information, much of which is considered "doomsday" talk, and yet these same people often hear reasons why they should not believe what they are hearing. Nobody wants to hear bad news, especially if they are not sure it is even true. So, life goes on as usual, and little effective action is taken.

As an economist and an ecologist, we do not consider ourselves experts in the debates over climate change. But we are, at the very least, interested and trained to collect, analyze, and evaluate information and data. Our fields and training intersect with the expertise involved. We are convinced the evidence of both climate change and its negative, cascading effects are real and may, in many ways, understate the dangers to our home planet and to the total quality of our collective lives—present and future.

The book you are holding, however, is not an attempt to convince anyone of the accuracy of the data concerning climate change. To be sure, it can be read as counsel to individuals and society in general, should the warnings be accurate. Although we make our evaluation of these issues clear, we think there is a much more salient perspective. In this sense, we are reminded of Pascal's Wager; namely, the logical construct of the Renaissance philosopher Blaise Pascal. In arguing for the existence of God, he contended that one should act as though God exists, because if He does, one gains everything, and if He does not, nothing is lost. On the other hand, if one assumes God does not exist, you either gain nothing if He does not or lose everything if He does. Assuming God exists, therefore, is a win–win situation, and assuming He does not is a lose–lose situation. The argument is obviously more about what you believe and how you act in *response* to that belief than about what is actually true or, in other words, the accurate state of nature.

Similarly, it is our position in this book, and a position we take in the strongest possible terms, that proceeding as though the environmental threats to the planet are completely true is by far the best course of action for human populations. The reason is not only because we are convinced the warnings are true, but also because this course of action—a positive, energetic movement toward social-environmental-economic sustainability—has the potential to create a much better and more prosperous future for people and communities than will blindly pursuing the current growth ethic, which is the central feature of virtually all public policy worldwide.

Parallel to the advancing crises in the natural world, we see a much more publicized meltdown of our economic system. The housing "bubble," accompanied by unsound regulatory and fiscal policies and reckless financial shenanigans on Wall Street, led to sharp drops in housing prices and nearly total collapse of our central financial system. Subsequently, problems in the financial sector, as they always will, spread to the real sector. The financial disruptions were quickly followed by widespread unemployment and a sharp recession. The current economic crisis in the United States is clearly the worst since the Great Depression of the 1930s, and no one can say how or when—or even whether—it will recover fully.

Are these two crises—as it were in the natural and the produced worlds—connected? We are convinced they are, and our work, as such, is largely dedicated to developing and explaining these relationships. This book may not seem on the surface to be about solutions. Nonetheless, we believe that anything qualifying as a "solution" demands an understanding of these connections. Our belief is that beginning with an understanding of how the real systems work is the only true path to effective solutions.

Before exploring the essentially methodological task of reconciling the *disciplines* of economics and ecology, however, our starting point needs to be reiterated: *The economy and the environment are in simultaneous crisis mode.* This is neither an academic nor philosophical point. It is an observable fact. And it is this fact that motivates us. If we, as a culture, do not respond effectively, the quality of life for our children, grandchildren, and all future generations will be dramatically impaired. To understand the cause of these interactive problems, we must beginning by reevaluating the modes of thinking that gave birth to the problems; because, in Einstein's widely cited view, the mindset—the level of consciousness of cause and effect—that created a problem in the first place cannot be employed to solve it. Changing behavior requires elevating the level of consciousness as a first order of the day. The present paradigm, which led us into the worldwide crisis and threatens to further engulf our world, is *not* the paradigm that can lead us out of it. We must, therefore, think and act anew.

Understanding the Inviolate Connection between Economy and Environment

In Greek, both *ecology* and *economy* have the same root, *oikos*, meaning "house." *Ecology* is the knowledge or understanding of the house, and *economy* is the management of that house—and it is the *same house*, one that cannot be divided against itself and remain standing. The modern world has too often forgotten or not understood this connection. As a biologist (CM) and an economist (RB), we seek through our collaboration to illuminate understanding, and in so doing to extract a semblance of guidance for individual behavior in these very confusing and perilous times. This task has both philosophical and methodological aspects, and we operate from a primary observation: namely, *the natural environment, of which we are an inseparable part, and the resource base for the economy are one and the same.*

This critical fact is often overlooked when addressing the economy–ecology connection. Management of the house is an inextricable part of the structure of the house. Too often, meeting the imperative of a robust economy, especially when the perceived need is incessant growth, has caused us to overlook the fact that there is no source of economic inputs or resources other than the natural environment. Any adjustment to this "natural" resource base causes systemic changes in the relationships among all the components. This is much like touching the corner of a waterbed—any contact whatsoever affects the whole of it. And those changes are irreversible, despite our overuse of the euphemistic terms *renewable resource* and *ecosystem restoration*.

In order to deal effectively with our social-environmental crisis, it is necessary, first and foremost, to reconcile the disciplines of *economics* and *ecology*. This sounds suspiciously like philosophers talking at a time when real-world action is called for; but it is not an idle, ivory tower observation. Systematic pursuit of a core philosophy has throughout history always marked the development of any significant cultural value. The forces and ideas behind our current problems are no exception. We have pursued economic goals—and in light of supposedly sound economic principles—that, in our observational judgment, are destined to lead to ruin, because they are in direct opposition to the central principles of ecology. We seek to unravel the ways in which supposedly sound, conventional economic thinking has led us astray, and reconcile the necessary interdisciplinary principles. This is the major task of Section II of this book.

Uniting Economics and Ecology

Since the Renaissance and the subsequent Scientific Revolution, the hallowed scientific method has become the accepted mode of thought for any respectable academic work in the sciences and social sciences. Value-free analysis is seen as the basis for any legitimacy, and a value-based approach is dismissed as mere opinion or speculation. Our respective academic fields are rife with subjective thinking. In reality, there is *no* objectivity in science because every question is subjective. In addition, science can only disprove something—never prove it. Direct observation is the only means of proof—the only viable fact.

To be sure, we have each performed much work within the accepted methodological boundaries of our fields and have a healthy respect for what can be learned and accomplished through the use of these techniques. But modern society has paid a high price for pure expertise. Academic fields have become more and more compartmentalized and intellectually isolate. Therefore, to be considered an expert in any field, it sometimes seems necessary to know more and more about less and less. And this can easily divorce learned individuals from an ability to deal with real-world problems of a *systemic* nature.

Science is the discovery and study of underlying physical and biophysical principles that define our world. Technology can be described as the application of these principles in the creation of useful human artifacts. Our linear, compartmentalized approach to science and technology has created many amazing processes and products, which have supported human populations in all corners of the globe. We are often blinded to indirect trade-offs of technological change by the siren song of some new device and what it might allow us to do or enjoy in the immediacy of the moment.

Economics, which is the study of how best to meet the material necessities of human beings, has willingly gone hand-in-glove with this march of science and technology. An economic imperative has often directed them. This alliance has encouraged individuals and societies to greet most new technological advances with collective, unquestioning open arms. The coining of the familiar phrase *labor-saving device* suggests that if one is blessed with some wondrous machine to assist in doing necessary work, then good things will happen.

Originally, these "good things," as Thomas Jefferson contended, were to be the freedom to pursue philosophy, the arts, and higher forms of culture. In modern times, this freedom, bolstered by the frantic drive to increase productivity, has often evolved simply to becoming even wealthier in the monetary sense. Again, stimulated by the *mis*application of basic economic principles, current thinking is spurred by the growth ethic corollary of "More Is Better." Growth has, therefore, become the dominant goal for both the public- and private-sector approaches to our macroeconomies, and thus has been allowed to take over both individual and collective actions. As ecology teaches us, permanent growth of any organism within a functioning ecosystem is a physical impossibility. This fact starkly identifies the dilemma facing the modern world.

Other Important Threads

The foregoing brings us to the major themes of the book. The linear, throughput-based treatment of nature's biophysical services, upon which economic activity depends, must be replaced by *systemic thinking* (as opposed to the current symptomatic thinking), which acknowledges that the environment and the economy are an ecosystemic unit and must be dealt with accordingly. We humans must begin to understand and accept that the environment will be forever changed by our efforts to produce goods and services in order to meet our desires, and that mitigating the effects of economic activity, as though the original status quo of the environment can be magically restored, is a biophysical impossibility and a human naïveté.

Always, we have an obligation to future generations. Our decisions and actions become the consequences with which they must live, and we give them no voice in our decisions. It is thus imperative for us both to understand and also to account for the ways in which those changes will occur as we extract resources, utilize the environment, manipulate the economy, or in any other way affect the resource base. There is, however, a caveat of responsibility: Is it too much to ask that we implement the precautionary principle, which places development projects on hold if the systemic changes appear untenable to the well-being of those generations? Modern approaches and policies appear to answer this question in the affirmative. In other words, we continue apace with economic activities designed to stimulate growth without understanding, or even seriously considering, the irreversible, long-term impacts our actions have on the natural-resource endowment of future generations. Moreover, they will not only inherit but must live with the consequences we bequeath them through our decisions and actions. Either way, they have no voice in that legacy, be it one of wisdom or of folly.

The growing danger with this cultural habit is twofold: (1) we are too certain of our knowledge; and (2) change is a constant process in which novelty and uncertainty reign supreme and forever. There are, however, growing differences of opinion as to what the impacts of a particular endeavor will be, as development projects have become larger and their impacts more global in nature. In the face of this uncertainty, the operating principle of our current economic paradigm has assumed the reckless route. We have allowed developments to proceed unless the impacts can be *proven disastrous*, as opposed to requiring effects to be *proven harmless* before allowing a development to proceed. Clearly, these are essentially inverse processes.

This is a fundamentally important point, and one that allows us to understand much of what goes on in our market-based, public-policy culture. Think about it. A range of possibilities exists with the proposal of a given development or the introduction of a potential product. If, for instance, there is deemed to be no substantial uncertainty about the unwanted effects of a potential product (i.e., widespread agreement), then the regulatory or permitting process becomes clear (e.g., DDT causes birth defects, and is banned, but after the fact). If the range of uncertainty is wide, then controversy surrounding what is allowed and not allowed is virtually certain to exist. Here, the fundamental point is that we are in control of a product or endeavor *before* it is introduced into the environment, but *once introduced, it is forever out of our control.*

It is easy to see that much of the widely debated controversy over government can be traced to just such situations. Activists call for increased regulation, while more laissez-faire types argue that the threats are overstated, and furthermore, government is innately ineffective and unable to do anything right. The strident disagreement that ensues is an ideal breeding ground for the rancorous politics and culture wars that have characterized U.S. society in recent times. At first glance, this can appear to be serious and intractable controversy, stemming from deeply held moral values, but can more often be traced to simple disagreements about how we should behave in the face of uncertainty.

Here, the intended role of science, through empirical data and research, is to better understand, and therefore reduce, the range of uncertainty. Ideally, this will decrease public controversy and make whatever turns out to be the appropriate regulatory rules and practices more acceptable. In practice, however, it has often not worked out that way, because there is an increased tendency to co-opt science in the defense of some desired policy or outcome.

Toward Socioeconomic Sustainability

In light of the foregoing discussion, the question is as follows: What should we do? A conventional answer to questions arising from disagreements over impacts, both short-term and long-term, is to call for a redoubling of efforts by scientists and engineers to measure those impacts in a bid to reduce the uncertainty and thus the controversy. Although we should not give up scientific quests for better information, finding a definitive answer is a utopian hope. We must always remember that such efforts, which rely on data and the scientific method, represent the kind of linear thinking that created the problem in the first place—Einstein's intellectual trap. Moreover, we will never have perfect knowledge and, thus, absolute certainty.

Sustainability is an ongoing journey of consciousness, not some mythical endpoint as some scientists, engineers, and politicians would lead us to believe. The journey must begin with the necessary philosophical and intellectual changes in the way we approach our entire culture, and those changes must be systematically applied to all that we do.

This is a big order; and we are aware that countless treatises have been presented over the last two decades invoking the overused term *sustainability*. At first, there was great questioning of the term when it arose in public forums. How is sustainability defined? What do you mean? Isn't it just environmentalism by another name? Isn't it merely another way of challenging corporations, a market-based economy, and the growth ethic? I (RB) have even heard it called "a communist plot to destroy capitalism."

In the last few years, however, the term has finally made its way into conventional usage. Politicians call for sustainability, and corporations announce they are promoting it. We are unimpressed because much of what appears to be happening, as has often occurred in our mass-media-dominated culture, is an overuse of the term that effectively renders it meaningless. In fact, we sense the term has come to be viewed by the powers that be as a harmless construct of the educated elite, and not anything that can pose a serious threat to efforts promoting vigorous economic growth.

After a careful, simultaneous examination of the current world environment and the global economy, we note that nothing less than a new sustainability-based worldview is required. This new worldview must have the potential to repair our cultural health and home planet in substantial measure. This is a serious challenge, but the problems are too severe for anything less.

Plan of the Book

Section I of the book seeks to establish the methodological and biophysical principles needed for the task of developing the concept of socioeconomic sustainability. The methodological observations of Chapter 1 may be seen as operating rules for ensuring consistency between the approaches of economics and ecology. Many of these have been raised in this Introduction, but much elaboration is in order, and examples will prove helpful. Throughout, our approach is to compare how we have operated in the past with our proposed new operating principles, and thus to suggest more appropriate and long-term, life-enhancing guidelines for directing where we should be headed and how to get there.

All human activity depends on an expenditure of energy, and there is no available "new" energy source other than the incoming endowment of solar radiation. Chapter 2 establishes these principles. The necessary perspective, in our opinion, goes well beyond our dependency on fossil fuel or the much-discussed, current and impending energy shortages of imperial modern industrial society. The application of these vital principles and perspectives to the economy is discussed in Chapter 3. Some pertinent history relating to the underestimated and misunderstood role of the energy sector in our economy is offered in Chapter 4.

Section II, as we have implied, is devoted to a critique of economics as it has been practiced. We devote much more attention to the misuse of economics in the service of what increasingly appears to be a ruinous pursuit of material wealth and expansion. We examine ways in which the discipline and the language of economics have been so altered over the years that the impact of economic thought extends well beyond the original intention of supporting efforts to best ensure a more secure and comfortable subsistence for people. The economics we need is one wherein the principles employed are virtually one with the inviolate, biophysical principles of ecology. A culture or society that refuses to adhere strictly and clearly to the biophysical principles, which define all viable systems, cannot long survive in a healthy state.

Finally, we combine these ideas into what we trust will display a sustainable worldview based on systemic thinking. Although we intend to extract some conclusions for personal and public behavior, we will do our best to keep this book from becoming a policy-based treatise. Rather, it is more about what and how we think than what we do. People who are adequately conscious of the biophysical principles and who think systemically will normally tend to do the right things. Please accept this book as it is intended—our small contribution to a healthy and rewarding future for those alive now and for those yet unborn.

Endnote

1. Émile Boutmy. *Études de droit constitutionnel* (Studies of Constitutional Law), Macmillan, London. 1891.

Section I

Setting the Stage

1

Methodological Overview

Any honest and compelling written or oral argument must begin with methodology. By that we mean that it should provide answers to such questions as the following: How did you reach that conclusion? What approach did you use? What assumptions did you make? How did you access disciplinary methodology or interpret available data? In this beginning chapter, we attempt to address such broad and important questions.

Although we are trained in economics (RB) and biology (CM), the methodology to which we refer goes far beyond our formal academic disciplines. We cannot abandon that training, but it must be used selectively, and in the broader context of rational inquiry. Much of the thrust of this book rests on the belief that conventional usage of the discipline of economics, and even some of the core tenets of the discipline, have led us astray. If this is true, then clearly, there must be some dominant methodological approach that would uncover these problems and perhaps lend guidance in charting an effective and appropriate path into the future. This section seeks to identify such an approach, and represents the search for a new mindset or worldview as much as it seeks new methodology. To accomplish this goal, we must compare and evaluate two disparate ways of thinking and their respective acuity of analysis: symptomatic and systemic.

Symptomatic Analysis

> Call a thing immoral or ugly, soul-destroying or a degradation of man, a peril to the peace of the world or to the well-being of future generations; as long as you have not shown it to be "uneconomic" you have not really questioned its right to exist, grow, and prosper.[1]
>
> **—E. F. Schumacher**

Conventional analysis is normally oriented toward treating symptoms, as opposed to uncovering the root causes for phenomena that are deemed problematic. The legacy of the Scientific Revolution, in the 17th and 18th centuries, weighs heavily in this regard. The hallowed scientific method has been the dominant explanatory and problem-solving mode of choice since the Enlightenment. This places heavy reliance on simple cause and effect.

For every action, there is supposedly an equal and opposite reaction. All phenomena are assumed to have direct causes, and if those direct causes can be uncovered, the problem is presumably on the way to being solved. Policy solutions, if solutions are sought, cannot be far behind.

For example, assume that unemployment is deemed to be a problem. Suppose further that the diagnosis of the problem is not enough jobs. This sounds logical—even obvious. Businesses create jobs; therefore, the superficial solution to the problem would be to give businesses effective incentives to create more jobs. This can lead to a focus on how companies are taxed and possibly a variety of other incentives that are known to exist.

The point here is not to propose a better way to cure unemployment, but rather to use this example to dissect the common approach to a problem. Recall that the initial diagnosis was "not enough jobs." Another plausible analysis of the problem might have been as follows: "The workers are not qualified." Indeed, there may be other reasonable explanations, but we will employ those two for our purposes. Note that a lack of jobs would immediately give rise to a focus on mechanisms to stimulate businesses, whereas a conclusion that the problem stems from unqualified workers might suggest that education and training are a more effective approach over time. Clearly, these are two very different strategies being employed in response to the perceived problem of unemployment.

The point is that both approaches stem from a cause-and-effect mindset whereby the analyst or observer says, "Here's a problem—what causes it?" On the surface, this sounds very direct, rational, and consistent with the tried-and-true scientific method. In practice, however, the assumption of a linear cause-and-effect relationship is likely not to yield effective results in a situation more complex than A causes B, B leads to C, C results in D, and so to fix D, we must work on A. The search for clear answers by assuming such a linear relationship can lead to complete ineffectiveness at best and serious mistakes at worst.

In the discipline of economics, which economists strive very hard to have considered a science, the mathematical and statistical models must necessarily seek to develop quantified functional relationships. If, in the real world, there is no consistent or predictable relationship between or among the variables in question (e.g., unemployment and education or unemployment and business incentives), then the models are worthless—or even counterproductive. Such an approach can cause us to ignore many other impinging factors under the false belief that a solution is "in there somewhere."

As an aside, it is often difficult for pure sciences to accommodate a lack of certainty and predictability. A person cannot be landed on the moon unless the scientists and engineers can predict exactly the effects of such things as thrust, weight, gravity, and so on. But economics is a *social* science, which means that human behavior is an integral part of the mix in developing and using the methodology that composes its core as an academic discipline. Other social sciences, such as political science, sociology, and anthropology, remember and honor this distinction. In fact, the occasional capriciousness of human behavior is embraced as a central component of each of these disciplines. They often delight in explaining how human beings are *not* rational, whereas economists stubbornly cling to the premise of the "Rational Economic Man," and they fret that if this overly simplistic, mechanistic assumption is not granted, the elegant mathematical models will lose their applicability and predictive powers. Of course, any devotion to methodology over the substantive topics and issues is a potentially dangerous mistake for any academic discipline.

Only economics, among the social sciences, is annually awarded a Nobel Prize. With tongue in cheek, it is noted that in seeking to deserve this award "for economic science," traditional economists take the restrictive assumptions that allow the preferred tools to be used a bit too seriously and arrogantly.

Economics places almost total emphasis on achieving efficiency. Efficiency, in the context of economic methodology, is the achievement of a given task with the least expenditure of resources. As one example, it amounts to a business creating a certain amount of product by using the fewest resources possible—and those resources are most likely to be measured as dollar expenditures, as opposed to physical amounts of materials or energy. This concept is customarily categorized as *efficiency in production.*

As a second example, a consumer is said to be operating efficiently if he or she is able to use a certain amount of goods with the least possible expenditure of income. To be sure, the income, or purchasing power, is representative of the effort, or the work necessary to secure that income through a job. Thus, minimum effort to achieve a maximum of use is, in economic theory, broadly defined as *efficiency in consumption.*

An additional wrinkle in the approach of economic theory is instructive at this juncture. In the production sector, the task can either be thought of as maximizing output subject to a fixed-cost constraint or as minimizing the cost of producing a fixed output. Similarly, in the household sector, the individual or family unit may either maximize use subject to a fixed-income (i.e., budget) constraint, or may minimize the necessary expenditure based on a fixed level of goods and services to be consumed or used. In either sector, these are inverse processes—one minimization, one maximization—and both are hailed as offering evidence of the flexibility of economic methodology.

However, neither of these optimization processes—whether referring to production for businesses or use for households—squares with the real world. To be sure, we always operate under constraints, but nature is effective, not efficient. These views of efficiency assume an independent variable to be maximized or optimized, (i.e., business output or consumer satisfaction) and then perhaps several dependent variables to be manipulated in achieving the efficiency goal. There neither is, nor can there be, an independent variable in any kind of relationship, whether natural or anthropomorphic, an observation that certainly includes economics. Thus, the supposed economic flexibility simply represents an attempt to intellectually manipulate the degree of dependence among variables. Under any circumstance, this amounts to a linear cause-and-effect methodological approach.

There are many fundamentally important issues revolving around the concept of economic efficiency. These will be explored further in Section II of this book.

Systemic Analysis

> I have no doubt that it is possible to give a new direction to technological development, a direction that shall lead it back to the real needs of man, and that also means: to the actual size of man. Man is small, and, therefore, small is beautiful.[2]
>
> **—E. F. Schumacher**

What is the alternative? If an overly scientific, linear, cause-and-effect throughput-based approach to the organization and management of human societies has led the modern world astray, what is the countervailing approach that would be more appropriate? The answer begins with systems theory and offers some first steps in how ecology can inform economics.

To begin with, the economic world must be viewed as a holistic, interactive system. A change in any variable affects everything else. The quantitative, statistically based approach, which has engulfed economic methodology in recent decades, initially purports to account for this holism, but economic practitioners in effect quickly despair of treating the bewildering complexity. Nevertheless, the instinct to preserve the methodology is powerful. The response, therefore, has become to engage in partial analysis (replete with the use of calculus and partial derivatives), whereby many factors are assumed constant—a physical impossibility in any relationship—while the dynamics among a few others is studied. Moreover, the analysis is often between just two variables, while the entire "rest of the world" is held constant. It is becoming daily more apparent that the unintended effects of such false assumptions are threatening our way of life.

Second, the real world operates cyclically, not linearly. Ecology can usefully inform economics in this regard. For example, populations of snowshoe hares in the boreal forest of Alaska and Canada increase in numbers of individuals for roughly 7 years before they crash. Meanwhile, the population of the lynx, which preys on the hares, increases as the hare population grows. When, however, the hare population crashes, the lynx switches its diet to grouse, which it rapidly depletes. During this time, however, the hare population is rebuilding and will once again become the lynx's staple diet—that is until the next crash, when the grouse takes over, and so on.

It is obvious to ecologists that changes in one component of a system set in motion forces that affect every other component—much like a waterbed in that you cannot touch any part of it without affecting the whole of it—the overarching "Waterbed Principle." Systems adjust and self-correct. Economics, with its reliance on cause-and-effect methodology and the growth ethic, has been slow to recognize this type of systemic recursiveness. Again, the unintended effects are enormous.

Third, the search for efficiency in the optimization of a tiny number of (perhaps even poorly chosen) variables has led to a lack of effectiveness in the operation of the overall system. For example, the quest for maximum profitability of business firms, or the search for maximum income and thus purchasing power of certain people, has dramatically reduced the capability of the overall economic system of ensuring the well-being of all people worldwide or of the long-term health of natural systems. Everywhere we see ecosystems (forests, grasslands, oceans) on the verge of collapse due to the unintended effects of exploitation for human use.

Finally, full accountability within a system is necessarily mandatory. Economists regularly give lip service to the need for a "full system accounting." By this is meant that all impacts of a given resource allocation must be acknowledged and measured to the extent possible. But here the fetish for quantitative measurement again creates a dangerous pitfall. The environment is difficult to price extrinsically, and often the only thing done is to acknowledge this difficulty, and then proceed as if the price were zero, which in effect discounts the birthright of all future generations. Thus, the analyst hides behind the assumption of *ceteris paribus* (other things being equal), and thus retreats into the misleading world of assumed, linear, cause-and-effect relationships, partial analysis, and of ignoring the unintended, but dangerously disruptive, effects that clearly exist.

Once again, we contend that these unintended effects, perpetrated by the current myopically linear economic view, are largely the cause of the global, social-environmental crises we currently observe. In summary, the push for economic growth to accommodate the completely unsustainable number of humans on the planet represents a symptomatic approach, which falls far short of addressing the cause of the problem. By analogy, this symptomatic approach is like going to your doctor because you do not feel well. In turn, your doctor tells you that you must get more exercise and lose 20 pounds. To which you respond, "Can't you just prescribe a pill? I don't want to change my lifestyle." If, on the other hand, we modern humans are to live with any measure of dignity and comfort, the symptomatic rationale embedded in contemporary economics must give way to a systemic approach that recognizes and accepts the reciprocal interactions among all aspects of social-environmental sustainability worldwide. Granted, this sounds like a daunting task, yet in our view it is paramount to human survival. Our hope is to encourage the acceptance of this challenge, and to make the case that a more life-enhancing path exists than the dismal course we are now following toward a predictable culmination of great suffering and widespread deprivation.

An Evolutionary View of America

All societies develop a cultural profile or ethic. This includes a set of traits, characteristics, and beliefs that define them as a people. Americans are clearly no exception. We take great pride, as politicians regularly remind us, in our "National Character." By and large, the American character was formed in the 18th and 19th centuries. The first significant event marking what was to become an incessant trend of Westward Expansion was the French and Indian War, which ended in 1763. This was essentially a conflict between the French and British over the question of movement into the Ohio Valley from the initial colonial settlement of the east coast of North America. It highlighted the strife over which of the largely European inhabitants would settle the continent and how.

The 19th century, beginning with the Louisiana Purchase in 1803; and the Lewis and Clark expedition over the following 3 years saw emergence of the notion of Manifest Destiny, the conquest of the supposedly limitless frontier—and perhaps above all, the beginnings of our cultural fascination with the phenomenon of *growth*. This time period, marked by the romantic notion of the unsettled American West, evolved into the Robber Baron era and the development of large-scale, natural-resource-based, capital-goods industries. In turn, emergence of these industries provided the infrastructure that supported the development of the United States as the industrial superpower of the 20th century.

The maturing of industrial society, the growth of consumerism, and the (at least temporary) rise of the middle-class marked the 20th century—a historical first. Throughout this maturation process, the United States maintained its reliance on continual, unfettered growth and the notion that technological progress was the answer to virtually all economically related challenges.

Indeed, growth and expansion, supported by incessant technological change, has come to mark not only our national identity as a people but also the overt public policy of virtually every major nation. Much of our discourse will focus on these issues, but for now we pause to reiterate one clear fact: *The Western World, having for some time been driven by the credo of growth and expansion, is in crisis mode.* The global circumstances created by this ethos are dire, and there is little agreement on the best way out of the mess. Therefore, it is time, as we now address the topic of intellectual methodology, to stop and take a broader view.

Resources were abundant during the time Americans were developing their cultural identity. Consequently, the predominant economic policies reflected the assumption that given a rich and seemingly empty continent free for the taking, the scarce resource was people. In other words, the constraining economic input was labor.

At this point, a critical qualification must be raised. We referred to the continent as "seemingly empty." In actuality, it was not. It was rich in natural ecosystems and a vibrant indigenous American culture. Nonetheless, European settlers behaved as though it was empty for the taking. The history of the domination and destruction of these native ecosystems—human and nonhuman—is a long, familiar, and dismal chapter in our national history.[3] Moreover, this repetitive human behavior contains lessons that are pertinent to our current economic malaise.

For our purposes, however, the dominant American character, and thus the typical mindset of its citizens, was formed during an era of voracious competition for the plentiful resources. As long as one is unconscious of a material value, one is free of its psychological grip. However, the instant one perceives a material value and anticipates possible material gain, that person also perceives the psychological fear of potential loss.

The larger and more immediate the prospects for material gain, the greater the urgency and the willingness to use political power to expedite the object's exploitation, because not to act is perceived as losing the opportunity to someone else. And it is this notion of loss that people fight so hard to avoid. In this sense, it is appropriate to think of resources as controlling people, rather than the reverse.[4]

Historically, then, any newly identified resource is inevitably overexploited, often to the point of collapse or extinction. Its overexploitation is based, first, on the perceived rights or entitlement of the exploiter to get his or her share before someone else does and, second, on the right or entitlement to protect his or her economic investment. There is more to it than this, however, because the concept of a healthy capitalistic system is one that is ever growing and expanding, but such a system is not biologically sustainable. With renewable natural resources, such nonsustainable exploitation is a *ratchet effect*, where to ratchet means to constantly, albeit unevenly, increase the rate of exploitation of a resource.[5]

The ratchet effect works as follows: During periods of relative economic stability, the rate of harvest of a given renewable resource, say timber or salmon, tends to stabilize at a level that economic theory predicts can be sustained through some scale of time. Such levels, however, are almost always excessive, because existing unknown and unpredictable ecological variables are converted, subconsciously if not overtly, into known and predictable economic constants in order to better calculate the expected return on a given investment from a sustained harvest.

Then comes a sequence of good years in the market, or in the availability of the resource, or both, and additional capital investments are encouraged in harvesting and processing because competitive economic growth is the root of capitalism. When conditions return to normal, or even below normal, the industry, having overinvested, appeals to the government for help because substantial economic capital is at stake. The government typically responds with direct or indirect subsidies, which only encourage continual overharvesting.

The ratchet effect is thus caused by unrestrained economic investment to increase short-term yields in good times and strong opposition to losing those yields in bad times. This opposition to losing yields means there is great resistance to using a resource in a biologically sustainable manner, because there is no predictability in yields and no guarantee of yield increases in the foreseeable future. In addition, our linear economic models of ever-increasing yield are built on the assumption that we can, in fact, have an economically sustained yield. This contrived concept fails in the face of the biological sustainability of a yield.

Then, because there is no mechanism in our linear economic models of ever-increasing yield, which allows for the uncertainties of ecological cycles and variability or for the inevitable decreases in yield during bad times, the long-term outcome is a heavily subsidized industry. Such an industry continually overharvests the resource on an artificially created, sustained-yield basis that is not biologically or ecologically sustainable.[6]

With the above in mind, American settlers could be likened to an invasive weed growing in a bare vacant lot. Thus, it is no surprise that American culture came to emphasize competition, private property (which is progressively extended to the realm of patents for everything imaginable, including ideas), and initiative, along with an ever-restless search for a new frontier to exploit.

All indications suggest this time has passed. The future environment, for the United States as well as all nations of the world, is no long one of superabundance for the taking. We have overpopulated the planet and gone to great lengths to exploit easily available resources. The low-hanging fruit has long ago been harvested. The world is telling us, in many ways and in no uncertain terms, that the environment we are in, and will continue to be in, is one of growing scarcity. Under such circumstances, the broader principles of evolution suggest that competition must yield to cooperation, growth must give way to biophysical sustainability, and consumption for its own sake must be supplanted by careful use of resources to meet basic human necessities.

These broad, sweeping generalizations deserve both critical examination and far greater detail. In summary, they suggest that the current economic crisis is well beyond a temporary malfunction of the dominant, socioeconomic institutions (e.g., the banking system, the housing sector), and rather underlies a full-blown crisis of culture. If so, the symptomatic thinking that got us into this calamity must change radically and quickly. The cultural ethos that spawned the problems cannot effectively get us out. We must therefore raise our cultural level of consciousness from symptomatic thinking to *systems thinking* if we are to extricate ourselves from our current dilemma and not merely pass it forward to future generations. This is a tall order for any society, no matter how adaptable or flexible it may be. This book is our attempt to assist that transition.

Lessons from Our Energy History

Setting the Stage

In October 1973, the Organization of Petroleum Exporting Countries (OPEC) oil embargo of the Western world ushered in what came to be known as Oil Shock, thus beginning what we in the United States called the "Energy Crisis." To the common observer, the media, and to most politicians and business policy people, the main indicating variable became rapidly rising prices for petroleum and petroleum-derivative products. Until then, prices of crude oil had been held between $2 and $3 per barrel for well over 10 years—from before 1960 into the early 1970s. Starting in 1973, however, prices began fitful increases, reaching a high of $34 per barrel in the very early 1980s.

Because oil provided at least 40 percent of the energy needs of the world, the resultant shock waves to the world economy were, not surprisingly, severe. Data for both unemployment and inflation rose into double-digit figures by the end of 1974. We were experiencing the dreaded *stagflation*, or the simultaneous presence of both unemployment and inflation, which Keynesian economic orthodoxy held could not happen. But, why not view the whole energy experience simply as a price shock, albeit to an important resource, which, nevertheless, would be overcome in due time with the typical involvement of technology?

The answer to this question lies squarely in the bailiwick of economic theory and revolves around the fundamental pervasiveness of energy as the primary resource, or driver, in all economic processes. Furthermore, the more complex and technologically advanced economic processes become, the more critical to society's economic well-being is its necessity of relying on available supplies of fossil fuel. As with every relationship, there is a trade-off; namely, with every increase in the reliance on fossil fuel, there is a mounting vulnerability of an economy or economic sector to price shocks.

As it turned out, there were many fundamental lessons to be learned, well beyond merely dealing with the price shocks to an industrialized-world economy. At the outset of Oil Shock, even professionals, who should have known better, did not completely understand the permeating role of energy—in particular, fossil fuels. Moreover, the tools we were accustomed to using when analyzing such situations were woefully inadequate for the task. An anecdote will best serve to underscore this point.

The Energy-Inflation Connection

In the heat of the initial inquiries during the onset of Oil Shock, and during the height of the stagflation experienced in 1974 to 1975, U.S. President Gerald Ford requested economists to estimate the proportion of the double-digit inflation that was energy based. Because empirical evidence revealed that direct purchases of energy averaged 8 percent of the total costs to or expenditures by U.S. firms, that figure (about one-twelfth of the perceived inflation) was initially hypothesized to be the answer. After all, it was reasoned that the other 92 percent of nonenergy expenditures would tend to shield a given producer from energy-related inflation.

The rate of inflation had moved from 4 percent in 1971 to as high as 12 percent during 1974. By applying this methodology to explain the increases, the indication was that only two-thirds of one percentage point could be ascribed to energy, which means that in the absence of energy price increases, we would still have experienced inflation of about 11.3 percent.

Subsequently, however, the tool of input–output analysis was employed to address this question in a different manner. Input–output is a more systemic analytical tool than most typical economic techniques and has the ability to recognize cumulative effects as materials move through an economy. It addresses how raw materials become intermediate goods and services and then final products by flowing from primary or raw-resource suppliers, to intermediate producers, and ultimately to final producers of goods and services. In short, it can follow the process of transformation of materials from extraction through final use. At each stage, value of the intermediate good is increased through the application of additional productive inputs, whether human labor, energy (solar or otherwise), recycled chemicals, or something else.

The estimated relationships between economic sectors (e.g., coal and steel, steel and automobiles, and so on) are determined by the state of technology and the underlying structure of the economy. As such, those relationships represent the percentages of all purchases by or sales by a particular economic sector going from or to any other sector.

In simpler terms, input–output analysis would recognize that the purchase of an intermediate good represents the purchase of previously expended energy, which the value added to the intermediate good represents. The logic is much like that supporting the original labor theory of value, as first developed by Adam Smith, David Ricardo, and Thomas Malthus. According to these Classical-School economists, the source of all value was labor, because human effort is indispensable in bringing forth the production of anything useful to humankind, and labor is therefore inherently represented in the purchase of any goods or services. Consequently, according to classical reasoning, the pricing of any good relies primarily on what it takes to recoup the monetary value of the energy expended in the production of the good—from beginning to end, human labor included.

The same analysis must be applied to energy. If, for example, General Motors buys a tire from Goodyear to put on a Chevrolet, part of the purchase price must be allocated to compensate for prior expenditures on energy by Goodyear, and indeed even earlier by the owner of the rubber plantation; the collection, processing, and shipping of the raw materials; the creation of the tire; shipping of the tire to the supplier; and thus to Goodyear, Chevrolet, and finally the new owner of the automobile. This is clearly the reality, even though the tire does not, on the surface, appear to represent a direct purchase of energy in all its various forms along the way to its new owner.

Although it is not our purpose here to critique the work—and the potential mistakes—of these early economists, one point is instructive. By assuming nothing is of value until human effort is expended on it, they essentially assumed the bounties of nature, such as forests, water, minerals, and the life-giving endowment of incoming sunlight, were free. Of course, they were groping for a rationale to affix pricing in a marketplace, and a so-called "free good" did not compute in the calculation of a fair price. This early ignoring of nature's services and their *essential, real value* to humankind exemplifies our critique of the "money-centeredness" of today's typical economic analysis. In fact, this early uncoupling of economics from the ecosystem services underpinning the flow of energy that drives the economy is a seminal error, which still contributes to the tendency of economists to think only in terms of how dollars flow, as opposed to the systemic flow of real energy—and of other valuable land-based resources—of which dollars are symbolic.

By using an input–output analysis, the new—and dramatically increased—estimate of inflation pointed out that the energy sector was responsible for about two-thirds of the inflation we were experiencing. In other words, in the absence of the price disruptions caused by Oil Shock, the rate of inflation, compared to the late 1960s, would have increased from 4 percent to between 6 percent and 7 percent as opposed to the 12 percent we experienced. Given that economists (in the language of the day) were ascribing some inflation to the delayed impacts of simultaneously waging Lyndon Johnson's war on poverty and the war in Vietnam, this would appear plausible. The Consumer Price Index was approaching that level by the precise time of the OPEC oil embargo in October 1973.

This recalculation of the role played by the energy crisis in explaining the highest historical rate of inflation in the United States to being responsible for about 65 percent of the impacts from an initial estimate of less than 10 percent indicates something more profound than just a slight underestimation. It suggests that the economists, like pretty much everyone else, did not understand the role of energy in our overall economic structure.

Two observations are significant. First, the tools customarily employed are overly linear, seeking simple, symptomatic cause and effect, and do not adequately represent a systemic point of view, which not only recognizes but also accounts for the importance of holistic relationships. Second, as was just mentioned, economists automatically focus on money, but not on the various streams of energy and materials.

As a result, energy had simply been treated like any other input, which in effect caused it to be virtually ignored historically. Here, standard economic analysis can help us out, because it can be applied to any situation in which prices are changing for some economic aspect of production. Producers substitute toward the use of relatively less expensive resources and away from those that are becoming relatively more expensive. This means that producers demanding energy in creating products or services will simply move toward a more conservative use of that increasingly expensive resource. Thus, at least over time, technological change is assumed to inexorably accommodate this adjustment process. Devotees of economics, whether economists or business leaders, supported by a faith in technological innovation and the pricing system innately tend to assume this process will go relatively smoothly. This being the case, economists did not expect or predict a crisis.

Unfortunately for any economy—especially a highly developed economy—there is no substitute for energy. It is needed ubiquitously in all productive processes. Because the massive technological changes, which might have been able to alleviate the situation, were impossible to obtain in the short run, a situation *temporarily* representing absolute scarcity was set up—for which the only safety valve was inflation. Consequently, confusion, marked by much finger-pointing and social-political discord, reigned supreme.

Toward Economic Reality

Some pricing information will help make the point here. Gasoline brought prices at the pump of 30 to 34 cents per gallon, even in the late 1940s. (Of course, the actual signs at the service stations would say 31.9 or 32.9 cents, but that is just marketing at work.) Over 25 years later, at the onset of Oil Shock in 1973, the *nominal* price (unadjusted for inflation) was still in the range of 32 to 36 cents per gallon. Economically, this long-term stability of the nominal price means that the *real* cost of gasoline, supported by the cost advantages obtained due to expansion in extraction, production, refining, and marketing, had dropped considerably during the quarter of a century postwar period. Until the 1930s, oil production was little more than an infant industry. In retrospect, the industrialized world, and especially the United States, was becoming seriously hooked on petroleum immediately after World War II.

The importance for economic adjustment is that, because some moderate inflation occurred over that time, relative to all other inputs, the real cost of energy was actually declining over that extensive period. Clearly, it became easy to take for granted the role of energy. Corporations, entrepreneurs, and certainly consumers did exactly what economic theory would predict; they opted for the relatively less expensive resource (in this case, fossil fuel) in virtually all processes of production. As a result, the American economy went on a monumental, energy-consumption binge. Of course, we did not fully recognize it as the binge that it was until the retrospection of the energy crisis brought it to the fore.

In addition to the extremely wasteful use of energy, there was a tremendous amount of technological change during that postwar period. The innovation process could afford to be anchored on and take for granted the readily available energy. Therefore, new technologies were incessantly steered in an energy-intensive direction, wherein they invariably expended more energy per unit of output than did the previous processes.

Further, since increased use of energy and the adoption of new, capital-based technology became almost synonymous, the low and falling real cost of energy stimulated many substitutions of fossil-fuel based machinery and processes for human labor (in economic theory terms, capital/labor substitutions). Indeed, the search for new ways to mechanize remains the favorite sport of American business today, although now the defining characteristic of that search also normally includes the involvement of computers and information technology.

The new, nagging problem during the hectic 1970s, and one that is only getting worse today, was unemployment. Without cheap energy to lavish on the job market, it has gotten increasingly difficult to maintain acceptable levels of employment. Accompanying that has been an unhealthy drift toward widening disparity in income and wealth. Increasing economic inequality, in our opinion, ranks as the primary problem threatening the well-being not only of our economy but also of our democracy. Considerable additional attention is given to this premise in Section II of this book, in conjunction with a more in-depth critique of the discipline of economics.

Of course, a rosy picture was painted for the employed labor force, even in light of nagging unemployment and growing income disparity. The increasing availability of capital led to empirically measurable increases in productivity, or, in economic terms, output per hour per worker. The articulation of the American Dream was again a setup, as it has been throughout history, to place the winners on a visible pedestal for public adoration, while either ignoring the losers or, even in some contorted manner, blaming them for the nagging problems that were arising and due to get even worse.

Understanding the Crisis

Even then, many wanted to ignore that message. Blame for the crisis was extended in virtually every direction: undemocratic oil "sheikdoms," profit-hungry oil monopolies, inept government energy regulations and policies, and so on. Little credence—but much informed denial—was placed either on our wasteful habits *or* on the possibility that we were beginning to run out of an important, finite resource and that the planet was sounding an early warning bell—however much or little time we might have left.

After all, the tasks of developing new sources of energy or technologies—and certainly the restructuring of an entire world economy—are obviously and unquestionably long-term propositions. It is entirely plausible that the price shocks, which began in the 1970s, could be interpreted as a forward-thinking, market-driven response to the onset of an era marked by increasing absolute scarcity of a vitally critical resource. This would have been a wise and helpful conclusion, and in fact just the sort of signal that a responsive pricing system is supposed to send. But this was not the collective interpretation—very possibly because very few *wanted* that to be the interpretation. The perceptions of and the implications for the short-term, vested interests were simply too numerous and too strong.

Consequently, the industrialized nations have indulged in an energy binge justified by more than three decades of political denial. The party has continued, marked by stubborn adherence to the premise that continued economic growth will somehow overcome the specter of absolute scarcity of air, land, water, and critical energy resources. Meanwhile, the flexibility for real-time adaptation is rapidly shrinking. Moreover, recent events in both the economic and environmental spheres clearly indicate that the necessary adjustments will be much more painful than they would have been had we heeded the initial "Limits to Growth" warning bells.

Regardless of the details of how Oil Shock occurred, or perhaps was engineered, the message conveyed by it and other environmentally related phenomena should have been clear: We may very possibly face a future wherein many natural resources—especially the marquee resource of petroleum—are in increasingly short supply. Technology can move the problem around, and shift the carrying-capacity stress to other resource systems, but it cannot simultaneously relieve pressure on all resources. Indeed, the trade-off of removing stress on one resource has the effect of inexorably increasing the stress on others.

Consider the following "vignettes" that are typical of what was often heard and said at the time, to which we have added editorial comments:

- *If oil is in short supply, we should use more coal.* This causes acid rain, causes global climate change, and puts pressure on water supplies and clean air, which ultimately pollutes the entire globe.
- *If food supplies are in question, we could bring more marginal lands into production and farm more chemically and intensively.* This threatens biodiversity, puts more pressure on forests, nonagricultural lands, water, and fossil fuel supplies used for fertilizer and pesticides, to say nothing of acidifying the world's oceans.
- *If acid rain due to combustion of coal and other fossil fuels threatens lakes and forests, we should develop new cleaner technologies and introduce new cleaner energy sources on a massive scale.* Of course, this invokes new and complex technologies, possibly with long lead times for

implementation, and requires other (probably hidden) energy subsidies from other existing sources, which themselves are not so environmentally benign.

- *If population pressures lead to urbanization and the conversion of farmland to urban uses, then return to the second vignette.* The entire line of analysis begins to resemble a recursive computer systems chart, except there is no escape from an infinite, self-reinforcing "do-loop."

Extracting the Meaning

Even though the list of vignettes that capture the conventional wisdom of the time is virtually endless, adding to that list is not our purpose. Rather, our point is that we were clearly thrust into the era of "everything is connected to everything else." We have always been in that era. It is simply the case that resource-related events during the tumultuous 1970s underscored that point and began to send what should have been unmistakable signals.

This glimpse into the history of events, beginning some 40 years ago, allows a practical overlay that dovetails with material to follow in the next chapter. Some significant general principles can be extracted:

- Energy use is inherent in all that we do—both in the creation of problems in the first place as well as the search for and implementation of potential solutions.
- Technology, even if it appears to be strikingly successful, merely shifts pressure from one resource system to others. We are left hoping that the newly pressured systems have considerably more carrying capacity than the old.
- The role of time is critical in two ways. First, the technological lead time must be short enough to avoid a serious crisis. Second, any symptomatic innovation may merely buy some time in any case. How long will a technological fix to a symptom last, when a systemic cure is required but ignored?
- Resource scarcities apply not only to quantities of *physical* resources, such as petroleum, minerals, or farmland, but are inextricably entwined with quality of *environmental* resources, such as clean air, clean water, ozone, and climate.
- Specifically, *water,* for which there is no substitute, may ultimately prove to be the limiting factor in the expansion of human activity.

These principles serve to complement and perhaps elaborate the laws of thermodynamics and the biophysical principles to be discussed subsequently. In some ways—and that is our point—they merely restate those fundamental laws. Nonetheless, Oil Shock and the ensuing energy crisis were merely the tip of the iceberg that ushered in the era of concern about perpetual and generalized resource scarcity. The collective point of these two chapters is that we must accept and abide by the most fundamental physical laws governing all activity on the planet. Any other path is rife with failure.

Proper attention to this entire collection of factors will represent no less than an alternative worldview. In the words of Thomas Kuhn, in his landmark book *Structure of Scientific Revolutions*, a new paradigm must emerge.[7] As with all the revolutions that Kuhn discusses, its emergence will not be without great controversy and resistance, because the required changes inevitably fly in the face of conventional wisdom.

In other words, cultural evolution expresses itself through changing values.

Paradigm Shift

Culture is not genetically inherited. It can only be learned from the past, modified in the present, and passed on to future generations. The notion of culture poses two questions: (1) What happens when the evolution of culture tears the social fabric with great force because of a shift in values in one part of society? and (2) How do we heal the social rupture that results from such a shift in cultural values?

Trying to answer these questions helps put the idea of a paradigm shift in context with our understanding of a profession as a microcosm of societal dynamics, such as forestry in the United States, which is relatively young, rich in experience, and *was* noble in its early vision. But the vision of its inception—once on the cutting edge of social responsibility, science, and correctness for its time—has dimmed and is rapidly being relegated to cultural history. Be that as it may, prior to casting out an old paradigm, wisdom dictates that we have a new one to take its place.

Each new paradigm is built on a shift of insight, a quantum leap of intuition, with only a modicum of hard, scientific data. Those who cling to the old way often demand irrefutable, scientific proof that change is needed, but such proof is seldom available to the diehard's satisfaction. Ironically, however, today's old way of thinking was yesterday's new way of thinking, which was challenged by an even older way of thinking to prove change was necessary or even desirable.

Time and human effort have proven the old paradigm to be more "correct" in terms of contemporary knowledge than its predecessor, but still only partially "correct." So it is with the new; it too will be more "correct" than the old *and* will eventually be proven to be only *partially* "correct," hence in need of change.

The personal and professional trap of every paradigm lies in its self-limiting nature, which manifests itself when the paradigm becomes too comfortable. At that point, new data cannot fit into the old way of thinking, which has grown rigid with tradition and hardened with age. It is thus necessary to periodically crack open an old belief system if a new thought-form is to enter and grow, moving both the individual and the profession forward in a renewed sense of authenticity in keeping with the cultural times.

Moving forward may be difficult for those whose belief system and personal identity are totally invested in the old paradigm, wherein their perception is vested in the cobwebs of the past, which preclude seeing any reason for change. For those who subscribe to a new paradigm, moving forward is easier, because there is something exciting and novel toward which to move—an opening vista that hints at what the profession must become, a vista more in tune with the knowledge and understanding of the day. Yet those who harbor the new ideas are not better as human beings just because their views differ from those who cling to the old patterns of thought.

The British historian Arnold Toynbee asked the critical question: "Why did 26 great civilizations fall?"[8] The answer, he concluded, was that the people would not, or believed they could not, change their way of thinking to meet the changing conditions of their world.

Thus, a profession can move forward only to the extent that individuals within the profession accept new philosophies and practices as demanded by a rapidly changing culture. No profession can remain the same. Those who feel they cannot accept new ideas must—and will—fall by the wayside. The constant evolution of culture decrees that every new paradigm will eventually be replaced by one more correct in terms of contemporary knowledge. And we must bear in mind that *now* is always a time of change, because change is a universal constant.[9]

This base prepares us to address the task of developing and understanding the concept of social-environmental sustainability. First, however, it is necessary to develop a more extensive critique of the theory and practice of economics as a discipline. This is the task of Section II of this book.

Endnotes

1. E. F. Schumacher. Thinkexist.com. http://thinkexist.com/quotation/call_a_thing_immoral_or_ugly-soul-destroying_or_a/296551.html (accessed on September 10, 2010).
2. E. F. Schumacher. *Small Is Beautiful: Economics as if People Mattered*. Blond and Briggs, London, UK. 1973.
3. Chris Maser. *Ecological Diversity in Sustainable Development: The Vital and Forgotten Dimension*. Lewis, Boca Raton, FL. 1999. 402 pp.
4. Donald Ludwig, Ray Hilborn, and Carl Walters. Uncertainty, resource exploitation, and conservation: lesson from history. *Science* 260 (1993):17, 36.
5. *Ibid*.
6. Chris Maser. *Resolving Environmental Conflict: Toward Sustainable Community Development*. St. Lucie Press, Delray Beach, FL. 1996. 200 pp.

7. Thomas Kuhn. *Structure of Scientific Revolutions*. University of Chicago Press, Chicago, IL. Second Edition. 1970.
8. Arnold J. Toynbee. *Civilization on Trial*. Oxford University Press, New York. 1948.
9. The foregoing discussion of paradigm shifts is based on: Chris Maser. *Our Forest Legacy: Today's Decisions, Tomorrow's Consequences*. Maisonneuve Press, Washington, DC. 2005.

2

Energy—The Critical Resource

The Flow of Energy Is the Only Real Economy

Economists are accustomed to thinking it is all about money. Our entire training is focused on measuring all real flows in terms of the dollar flows that *represent* them. In short, dollars are merely symbols of the "stuff of real value" circulating in the economy—they are not the actual wealth. The wealth (goods, services, energy, and human labor) moves in one direction while the money to compensate for the transfer of the wealth moves in the other. Economics, as it is structured, often induces us to misplace our attention on the money rather than on the item of real value.

Examples are available anywhere we care to look. If labor is hired to grow food, we note the wage rate or the total cost of that human input in producing the final product. If a good is transported, the shipping cost in dollars is of concern. If a resource, such as a mineral, is extracted for use in production, the question asked will be, "How much did it cost to mine it?" If a product is purchased at Walmart, the important number is supposedly the (dollar measured) price tag. In each case, economic analysis tends to ignore the real effort of humans, the real energy expended, and the real material substances extracted, converted, transferred, or used.

This dollar myopia can lead us to tragically miss the point of what is really happening in the economy. *All economic activity is a conversion of energy.* Without some type of energy flow, no economic activity can occur. In thinking about business and economic activity, we have become accustomed to a money-centered definition of what is occurring, and this is certainly true for the concept of economic efficiency. Something is *efficient* if it is provided at minimum dollar cost. It is termed *inefficient* if the dollar cost is higher than economists think it should or could be. As we face the challenge of moving toward socioeconomic sustainability, an alternative interpretation of *efficiency,* based on the flow of energy, will serve us much better. The entire concept of efficiency, steeped as it has become in the notion of minimizing dollar costs, should appropriately give way to effectiveness. Nature is not efficient—it is effective. To understand what is meant by *efficiency,* versus *effectiveness,* let's consider two examples: pine trees and the water supply of a small hometown.

Pine trees (as well as the thousands of other species of wind-pollinated plants worldwide) cast upon the winds of fortune a prodigious amount of pollen to be blown hither and yon. We say "the winds of fortune" because it takes an inordinate amount of pollen riding the vagaries of air currents to come in contact with and fertilize enough pine seeds to keep the species viable through time. Although an extremely *inefficient* mode of pollination in that many, many more grains of pollen are produced than are used to fertilize the available pine seeds, the system is highly *effective*, as evidenced by the persistence of pine trees through the ages. And if you are wondering what happens to all the "unneeded" grains of pollen, they are eaten by a variety of organisms, which benefit from an extremely rich source of nutriment. *Nothing in nature is wasted.* "Waste," as people think of it, is an *economic concept*—not an ecological one.

Now let's consider the water supply of the small town. As its computer network was being constructed, the town officials, like businesses and communities everywhere, were increasingly focused on all conceivable aspects of *efficiency* in order to eliminate as much perceived redundancy as possible in everyday activities, because they were—and still are—seen as a waste of money. Accordingly, everything in the town that could be computerized was computerized, to eliminate the *unwanted redundancy* of manual control—everything, that is, except the water supply, which was fortuitously overlooked.

Today, should the computer program that controls the water supply suddenly fail, the townspeople would still have water because the unwanted redundancy is really a *backup system* of manual override. The ability to manually override a computer failure (while considered *inefficient*) is *effective* in giving the town the resilience to overcome a potentially disastrous circumstance and remain viable, while other—more efficient—communities would not be so fortunate. Here, it is important to understand that no system has redundancies, only *backups*.

To illustrate, you were born with a built-in, backup system called a nose with two nostrils and a mouth. If, for example, one nostril becomes plugged, you can still breathe through the other, which gives you two options (a nostril and your mouth), depending on which nostril is stuffed up. When, as sometimes happens, both nostrils become plugged, you can breathe through your mouth, giving you a three-way backup system, which, barring a catastrophic event, gets you through life with relative comfort. The question, therefore, is which two of the three openings through which you breathe are *redundant*? In other words, which two could you live without—both nostrils or your mouth and one nostril?

With respect to ecosystems, each contains built-in, backup systems, in that they contain more than one species that can perform similar functions. Such backups give an ecosystem the resilience either to resist change or to bounce back after disturbance. Backup systems, in the biological sense, are composed of the various functions of different species that act as an environmental insurance policy—and thus a simultaneous economic insurance policy by ensuring a continual flow of true energy. Again, the flow of dollars is merely symbolic.

Follow the Energy—Not the Money

Instead of measuring the dollar cost of producing food, we must take note of the energy costs, both quantity and quality, of producing a particular amount of food calories. Rather than worrying about the shipping costs of moving goods around, we ought to note the necessary units of energy, both direct and indirect. Instead of simply focusing on the price tag at Walmart, there must be concern about the amount and type of energy required to produce the product and get it to the shelves of the store.

This notion of an economy as energy flows rather than dollar flows has profound, even revolutionary consequences—not just for understanding the modern economy, but perhaps even for the survival of our species. At the very least, understanding the current economic-environmental malaise demands that we explore the implications of these ideas in much greater depth. This chapter is devoted to that exploration, but in order to ensure the proper depth of this effort, we must begin with the assertion of some essential first principles.

Lessons from the Laws of Thermodynamics

The *first law of thermodynamics* states that the total amount of energy in the universe is constant. However, it can be transformed from one form to another. This means that the amount of energy in the universe overall remains the same, even if the passage of time were allowed to vary. Moreover, the existing energy would remain constant even if we went forward or backward in time.

This has many implications for our purposes. For instance, the commonly heard phrases of either "energy production" or "energy consumption" are basically *non sequiturs*. Energy is neither produced nor consumed in the overall sense. It is merely transformed. We are accustomed to imagining processes being in danger of running out of energy, or of momentarily depleting a particular source of energy. In effect, what is really happening is that a subsystem within which the process is operating may be experiencing a diminution of available energy, but the sources outside that subsystem need to be examined. Although the technological challenges for accomplishing this may seem daunting (e.g., how can we supplement with sunlight if we run out of oil?), the knowledge and insight gained by examining the larger system can have a calming, enlightening, and focusing effect for those responsible for an "even" flow of energy for public use. There is always a larger system—almost certainly involving solar radiation—whereby a search for new available sources of energy can be productive.

The *second law of thermodynamics*, or the law of entropy, states that the amount of energy in forms available to do useful work can only diminish over time. The loss of available energy to perform certain tasks thus represents a diminishing capacity to maintain order with a certain configuration of materials (say a tree), and so increases disorder or entropy. This disorder ultimately represents the continuum of change and novelty—the manifestation of a different, simpler configuration or order, such as the remaining ashes from the tree when it is burned. At first glance, it appears as though the ability to do work in that subsystem is done. In the bigger picture, however, the injection of solar energy over time (from a larger system—a star) can restore the ability of the subsystem (the forest within which the tree existed) to do more work. In turn, the ashes offer a reinvestment of biological capital in the form of nutrients to the soil and thus to the forest.

Finally, the *law of maximum entropy production* says that, "a system will select the path or assemblage of paths out of all available paths that minimizes the potential or maximizes the entropy at the fastest rate given the existing constraints."[1] The essence of the maximum entropy law simply means that when any kind of constraint is removed, the flow of energy from a complex form to a simpler form speeds up to the maximum allowed by the relaxed constraint.[2] A hot cup of coffee left sitting in a room will evolve to room temperature in a fairly short time period. If, however, the coffee is in an insulated cup (a constraint), it will take a little longer, but it will still cool. Clearly, we are all familiar with the fact that our body loses heat in cold weather, but our sense of heat lost increases exponentially when windchill is factored into the equation because our clothing has ceased to be as effective a barrier to the cold, or constraint to the loss of heat, as it was before the wind became an issue. (Of course, in the long run, an innovative solution would be to generate energy from the wind, and, in effect, tap into the larger system.) Moreover, the stronger and colder the wind, the faster our body loses its heat—the maximum entropy of our body's energy whereby we stay warm. If the loss of body heat to the windchill is not constrained, hypothermia and death ensue, along with the beginnings of bodily decomposition—reorganization from the complex structure and function toward a simpler structure and function. To avoid these dire consequences, we must remove ourselves to another system (go inside) or provide additional constraints (dress even more warmly).

The laws of thermodynamics are, in a sense, analogous to the Constitution of the United States, a central covenant that informs the subservient courts of each state about the acceptability of its governing biophysical principles. In turn, the biophysical principles represent the state's constitution, which instructs the citizens as to what acceptable behavior is within the state. In this way, nature's rules of engagement inform society of the acceptable, behavioral latitude (read "biophysical constraints") of its citizens if they want to survive in a sustainable manner. These principles are as follows:

1. Everything in the universe is a relationship supporting relationships, thus precluding the existence of an independent variable, absolute freedom, or a constant value beyond the number one—the universal common denominator; everything else is a multiple of one.
2. All relationships are productive, because all relationships produce an outcome.
3. The only true investment of energy on Earth is solar energy from sunlight; everything else is a reinvestment of existing energy—including all market dynamics, such as the energy that drives the stock market.
4. All systems are defined by their function—not their pieces in isolation of one another.
5. All relationships result in a transfer of energy, which is all we humans ever do.
6. All relationships are self-reinforcing feedback loops; whether feedback loops are positive or negative depends on whether or not they satisfy a human desire.
7. All relationships have one or more trade-offs.
8. Change is a constant process that produces only novel outcomes.
9. All relationships are irreversible because all outcomes are novel, and we cannot go back in time to recapture a past outcome.
10. All systems are based on composition, structure, and function, where composition determines structure, and structure allows function.
11. All systems have cumulative effects that compound unnoticed through time until a visible threshold is crossed, which makes them apparent.
12. All systems are open to cosmic energy; consequently, closed loop of anything (be it an economic system or technology) is a physical impossibility, because while energy can be constrained, it cannot be contained.
13. All systems function in curvilinear cycles, like a coiled spring, wherein each cycle approximates—but only approximates—its neighbor through the ever-changing process of novel outcomes, but each cycle is simultaneously linear because the coils never touch.
14. Systemic change is based on self-organized criticality, which simply means that an internal shift in one or more components of the system ultimately caused a dramatic shift of systemic proportions.
15. Dynamic disequilibrium rules all systems, which negates the romantic notion of the "balance of nature."

It ought to be clear by now that virtually all familiar processes could be used as examples, because the above biophysical laws that oversee everything we do as humans govern them. To our benefit, the laws of thermodynamics, including maximum entropy and the subordinate biophysical principles, can serve as a "language" in depicting the trade-offs of all our situations, actions, and behavior in terms of systemic sustainability. In essence, we are suggesting that such an interpretation of economic behavior would serve our culture very well. Such an interpretation is virtually mandatory if we are to approach a high level of socioeconomic sustainability.

In summary, systems are by nature dissipative structures that release energy by various means, but inevitably by the quickest means possible. For example, as a young forest grows old, it converts energy from the Sun into living tissue, which ultimately dies and accumulates as organic debris on the forest floor. There, through decomposition, the organic debris releases the energy stored in its dead tissue. Of course, rates of decomposition vary. A leaf rots quickly and releases its stored energy rapidly. Wood, on the other hand, generally rots more slowly, often over centuries in moist environments. As wood accumulates, so does energy stored in its fibers. Before the suppression of fires, they burned frequently enough to generally control the amount of energy stored in accumulating dead wood by burning it up. These low-intensity fires protected a forest for decades, even centuries, from a catastrophic, killing fire. In this sense, a forest equates to a dissipative system in that energy acquired from the Sun is released through the fastest means possible, be it gradually through decomposition or rapidly, depending on the amount of wood, the weather, and the resulting intensity of the fire. The ultimate constraint to the rate of entropic maximization, however, is the immediate weather in the short-term (hot and dry versus cold and wet) and the overall climate in the long-term, as in global warming, which is influenced by our attempts to circumvent nature's biophysical principles—in other words, to cheat for monetary gain and the power that comes with it.

Five Operating Principles

We conclude this chapter and Section I of this book by asserting five basic principles that will serve throughout the book to facilitate the discussion and exploration of the all-important links between economics and energy. Examples are incorporated with the principles in order to cement a better, practical understanding of this vital relationship. In their simplest form, these principles refer to (1) the real role of energy in the economy, (2) the need to count everything, (3) the need to focus on the quality of the energy, (4) the need to promote a diversity of sources, and finally, (5) the need to work *effectively* with nature in capturing the many bounties that are available.

Principle 1—The Real Economics of Energy

All economic activity constrains energy. It does not use energy.

All economic activity is a process of constraining the flow of energy to maximize its usefulness, while simultaneously conserving its sustainability and the right of future generations to the same quality of life we, in the present generation, enjoy.

Entropy, according to the laws of thermodynamics, is basically the ability to do work. High or low concentrations of energy within energy-providing materials represent the potential ability to perform useful tasks. Ability to accomplish work, in turn, is dependent on concentrations of energy-yielding materials. A piece of firewood contains concentrated solar energy from the last few decades. A lump of coal has within concentrated solar power from millions of years ago. Similarly, a gallon of oil or gasoline, termed *fossil fuel*, is probably more appropriately recognized as partial solar energy from bygone geological eras. It is nonrenewable because we cannot bring back sunlight, the plants that captured it, or the life-forms that used it from past millennia.

It is readily understandable that humans focused economic activity on such attractive sources of energy as firewood, coal, and liquid fuel, because they represent highly concentrated forms of energy. Burning them releases the stored energy, even as it reduces the material to ashes, carbon dioxide, and water, thereby offering the ability to do what we humans deem useful work. It can heat our homes, generate electricity, or power our engines. Of course, once the materials are burned, the concentration of energy dissipates, and the overall system within which the economic activity occurs moves toward a state of less available energy to do further work.

The process of accumulating sources of energy can be examined within this context. Because a situation of complete dissipation of stored energy would entail no concentrated ability whatsoever to accomplish work of any kind, the process of accumulating energy is essentially an act of concentration. We mine coal. We cut trees for firewood. We drill for oil the world over, which incites wars and much political strife—but that is another story. These are merely acts whereby we attempt to replace situations of dwindling concentrations of energy with those of high concentrations of energy.

Many of the common attitudes about this process are incorrect or illusory when examined in the larger picture. The fundamentally important point of the laws of thermodynamics is that we have no ability to create new energy but merely to move it around. Even as we seek to exploit and thus accumulate an energy source, such as mining a pile of coal or cutting a stack of firewood, the larger system within which we operate is moved inexorably toward the rapid dissipation of available energy. We may be pleased that we have stored some firewood for the winter, but the entire forest system is moved slightly toward a lesser ability to supply more firewood. The coal may have come from a mountain in West Virginia (a geological storehouse of a sort), but after the mountain is strip-mined, it resembles more of a flat plain that now is devoid of coal. And these examples refer merely to the extracting process, or the *obtaining* of an energy source, and do not address the entropic or environmental implications of the use or *transformation* (*not consumption*) of the energy in doing the work.

A major reason that the act of accumulating a concentration of potential energy moves us toward a situation in which the systemic energy is dissipated in the larger system is due to the simple fact that it takes energy to exploit a potential energy source. And this leads to a second principle.

Principle 2—Counting Everything

Accounting for the expenditure of energy must be focused on net energy and must include all direct and indirect costs.

Counting everything is a monumentally important, but often ignored, principle. Major miscalculations have occurred through failure to take the whole into account. The total cost of the energy used to drive the economic system must be accounted for and monitored in order to conserve biophysical sustainability and thus the productivity capacity of the total environment for all generations.

For example, during the energy crisis of the 1970s, oil shale was touted as a promising potential source. Huge deposits of oil-bearing rock formations, largely around Grand Junction, Colorado, are present in the United States. Unfortunately, these expectations ignored the realities of the net energy required in the extraction of the oil.

Subsequent energy-based analysis of five proposed technologies for extracting the oil revealed that only one even yielded a positive net energy—and that was by 1 percent. It would have taken 99 barrels of oil as an energy input to extract 100 barrels of final product. The other four technologies required more energy in extraction than the energy to be obtained by the process.

This episode reveals another common mistake in ignoring the reality of net energy in favor of dollars. Some energy officials at the time contended that the price of crude oil, then around $10 per barrel, was inadequate to make oil shale "economically feasible," but that feasibility would occur when the price of oil reached $15 per barrel. Apparently, they were projecting the cost of their required energy for extracting the oil at a constant level, even while assuming their revenues for the same resource would rise. This bit of irrationality ignores the fact that in pure economic parlance, their cost curves would rise at 99 percent of the rate of their revenue curves. A net energy analysis revealed that a more likely "economically feasible price" would be in the neighborhood of $600 per barrel, the lion's share of which would obviously be to purchase energy. Needless to say, we have yet to exploit oil shale as a source of energy.

Another popular misconception can be explained through net energy analysis. It is common to speak of a given "year's supply" of energy reserves, particularly of nonrenewable fossil fuels, such as coal, oil, and gas. Normally, these are simple gross calculations that ignore the costs of extracting, concentrating, storing, and delivering the energy in a usable form. For instance, if a resource were to be 20 percent efficient, in that it required 80 units of energy to extract 100 units for final use, then the supply available for use as a final product would only last one fifth as long as projected by a gross analysis. Thus, mistakes can be made by thinking in years as well as in dollars, as opposed to units of energy. We will return repeatedly to the concept of net energy but now will consider a third principle.

Principle 3—Energy Quality Is Vital

The focus must be on the quality of energy—not the quantity.

In times of energy abundance, the premium is on rapid growth, but when available sources become scarce, the survival advantage shifts to achieving high diversity within and among systems and organisms. Those components will be favored which use energy more efficiently and thus can adapt to a more stable environment, at least for a time.

When there are untapped sources of energy available, the competitive edge in the game of survival is in favor of organisms that are capable of rapid growth, even though they may seem to be inefficient. Observe a vacant and recently cleared lot, and one will typically see large weeds (often annuals) spring up the first year. Although not customarily long lasting, they are efficient in capturing and using the abundant solar energy falling on the lot, which they effectively harness for rapid growth.

In subsequent years or successions, however, the rapid-growth organisms will give way to a more stable and diversified suite of participants. These will usually be longer-living species, such as grasses or mosses, which use the steady flow of available energy more effectively over time than could the original, fast-growth specialists.

Ecosystems involving human communities are no different. When there are available, untapped resources, the premium within ecosystems will be on individuals and institutions that promote rapid growth. Competition is the dominant operational mode and the order of the day. Consider the United States in the 19th century. The (supposedly) untapped Western frontier was inviting territory for competitive, growth-oriented people, attitudes, and even legal structures. The learned behaviors emphasizing dominance of frontiers and American Manifest Destiny carried over into the 20th century, which saw a rapid, but occasionally unnoticed, filling of the planet.

Now, as we enter the 21st century, few can deny that we have reached an era of limits—an era from which there is no ready escape. Most reputable population biologists place the Earth's carrying capacity, with any kind of reasonable quality of livelihood (certainly below that of the average American), at no more than two billion people. Without further elaboration, it must be noted that the world population of humans is rapidly approaching seven billion.

Unfortunately, most economists were trained with an eye toward promoting rapid growth. The last two to three centuries have been powered by fossil fuels, which have given a historically brief spurt to something we now call the *growth ethic* through the use of limited supplies of energy accumulated over millennia of geologic time. Aided by this history-induced myopia, analysts and policy makers have given scant attention to systems that might realistically encompass sustainability, despite the fact that the longer version of history has approximated exactly that. Unfortunately, the geologically instantaneous fossil fuel era has obscured that fact. The days of the 19th century "cowboy economy" are over, and our intellectual constructs must be retailored to fit that reality. This leads us to another basic principle.

Principle 4—Promote Diversity

A diversity of sources must necessarily match a diversity of uses.

Promoting diversity is a subtle principle with far-reaching implications. Sources, which are, in reality, now marginal or rapidly approaching marginal, are often supported by hidden fossil fuel subsidies. When those subsidies are no longer available, such sources will not yield the net value they now appear to have. Some may become completely nonviable. As with the oil-shale example, economists often get this backwards. Marginal sources are seen as becoming viable with increasing scarcity and higher prices of currently used sources of energy, but a true net energy analysis might indicate they should be ignored completely.

On the other hand, energy sources and technologies, which are currently seen as marginal or at least incapable of producing enough power to merit attention, may well come into their own for specialized uses that are uniquely fit to their capabilities. The call for diversity in this principle is an important point. In a natural ecosystem, the succession of species will include a variety of organisms that behaviorally partition the habitat in myriad manners that allow the use of available energy in different ways. Thus, a rich diversity of species will characterize a state in which there is little or no untapped energy. The important lesson for us humans is that cooperation must replace competition, and diversity should replace "monoculture" as the order of the (new) day.

In human settlements, this might involve someone tapping wind power in appropriate places, farms using small-scale hydro, methane from livestock feces, selected use of solar panels, or even some combination of these for specific or remote purposes. The result will be an optimization of total power available to do useful work in the system, and will be facilitated by people with different skills, different geographical situations, varying financial capabilities, and different requirements for energy. Again, diversity and cooperation will be key to survival if any kind of high-quality lifestyle is to be achieved.

The role of technology, as always, deserves some special attention. In the Western world, reliance on a technological fix is legendary. In the United States, our culture is steeped with terms such as "American know-how" and "Yankee ingenuity." We have become accustomed to technology bailing us out of some perceived problem at the last minute, whether the problem is one of some necessary new process, a shortage, or even the need to win a war. Oil was discovered as whales were becoming depleted. Artificial fibers, such as nylon, were invented as the German U-boats made natural rubber for tires unavailable during World War II. We marvel that the new substitutes are often better than the old products or processes they replaced.

To economists, new technologies have tended to play the role of saving the economy from what they would call "secular stagnation," which is merely another term for *perpetual recession*. New processes create new jobs, first in the research-and-development phase, then in capital goods industries, and finally in the creation of new products, which supposedly stimulate a renewed spate of consumer spending.

To a critical eye, this indicates a worrisome and unhealthy trend. Historically, economic output was for the purpose of providing the means to meet our material requirements as human beings. In recent times, however, the need, often politically driven, has virtually become the means whereby the superficially apparent health of a mass economy is maintained. Instead of small-scale producers meeting the requirements of customers, people are stimulated to keep purchasing and thus meet the desires of large-scale producers—who are mistakenly seen as the "backbone" of the economy.

Additionally, energy technologies must be considered with great care. They may prove to be more of a trap than the apparent answer to all of our problems. This leads us to a final principle.

Principle 5—Work with Nature

Freely available solar radiation must be captured, stored, and employed—not ignored, purposefully disregarded, or replaced.

Natural flows of energy are available through sun, wind, water, and so on, which operate in the world without monetary payment. The successful economy of the future must maximize the use of, and not destroy, these free gifts of nature. They all have a monetary cost to develop, but sunlight not only is the sole true investment of energy in our world but also is the one source that is free of environmental costs—that is, until climate change exacts a cost even economists will admit to.

Examples abound as to how we have violated this principle in the past, and as to how we could better employ it. Technology always involves a transformation of energy. An automobile transforms fossil fuels into mechanical movement. The "work" of the movement is no more than 5 to 10 percent efficient, compared to the potential energy of the fossil fuel. It is testimony to our apparently incredible desire to get somewhere that we put up with—and ignore—such inefficiency, as well as the financial, environmental, and social costs.

In general, all transformations are inefficient due to friction or the innate nature of the process of transforming one type of energy to another (e.g., chemical to mechanical), wherein each involves a loss of potential ability to do work. There is *still* no such thing as a perpetual motion machine. We often pay a stiff, energetic price for getting the type or quality of energy we want for a particular job. Consider an example that also applies directly to Principle 4: a flashlight battery delivers a miniscule fraction of the usable energy that it takes to produce it. But the convenience and portability, when you need it, apparently make it all worthwhile. If the amount of electricity used during the normal life of a flashlight battery was employed productively *and* available from the power grid, it would cost less than 1 cent. But, because the flashlight battery does not cost a great deal, it represents one more way we are induced to think of our artifacts in terms of dollars, rather than the flow of energy they represent.

In summary, however, complicated transformations of energy through the use of involved or piggybacked technologies normally result in substantial inefficiencies, which means that the energy costs of modern processes are usually quite high. Ironically, we are accustomed to thinking of modern industrialized economies as efficient, and the more backward appearing economies of nonindustrialized nations as inefficient. Although it may seem counterintuitive, the opposite is the case with regard to the uses of energy per unit of output in these nonindustrialized nations. The so-called "backward" economies obtain a much higher output per unit of energy employed than do industrialized economies, such as that of the United States.

The problem is that nonindustrialized economies have so little access to energy resources to begin with. Actually, many nonindustrialized nations have abundant indigenous sources but have lost control of their own energy resources to foreign governments or multinational corporations simply because they lack available development capital. On the other hand, we in the industrialized nations have become so accustomed to such extravagantly available supplies that virtually all processes (think "labor-saving devices") have experienced capital or labor substitutions. These normally take the form of automation to eliminate as many workers as possible, until the production or manufacturing processes in question have become very energy intensive.

This means that each worker has a great deal of capital (hence, energy) with which to work and thus appears to be highly productive. Even though we enjoy higher wages, this trend ends up being energy wasteful and employs progressively fewer people.

For a future that promises to be marked by energy shortages and chronic unemployment, this current trajectory borders on insanity. The implications of this questionable and highly unstable situation are explored in more depth, in conjunction with the critiques of pure economic theory, elsewhere in this book.

There are many profound, economic implications of this seminal chapter, and we have only begun to touch on them. Many others are explored throughout the book. Suffice it to say at this point that the realities and usage patterns of energy are at the heart of both the economic and environmental challenges we currently face. No resource is more fundamental to operating a healthy economy or maintaining a livable environment than energy. Energy conversion is the necessary starting point for both economic activity and environmental sustainability. Both production and pollution begin with the garnering and conversion of energy. There is no hope for a transition, smooth or otherwise, to either socioeconomic or social-environmental sustainability without wise and thoughtful responses to our self-created energy dilemma.

Endnotes

1. Rod Swenson. Emergent Evolution and the Global Attractor: The Evolutionary Epistemology of Entropy Production Maximization. *Proceedings of the 33rd Annual Meeting of The International Society for the Systems Sciences*, P. Leddington (ed). 33(3):46–53. 1989; Rod Swenson. Order, Evolution, and Natural Law: Fundamental Relations in Complex System Theory. In: *Cybernetics and Applied Systems*, C. Negoita (ed.), 125–148. New York: Marcel Dekker. 1991.
2. Rod Swenson and Michael T. Turvey. Thermodynamic reasons for perception-action cycles. *Ecological Psychology* 3 (1991):317–348.

Section II

Economics in Theory and Practice

I believe that the great part of miseries of mankind are brought upon them by false estimates they have made of the value of things.

Benjamin Franklin (1706–1790)

3

The Innate Nature of Economics

It should be abundantly clear after Section I of this book that we, the authors, have serious criticisms not only of economics as a discipline but also of the prevailing economic thinking and perspectives as the economy has evolved. The next several chapters explore these perspectives in much greater depth.

It is appropriately symbolic that, as coauthors, we are an economist (RB) and a biologist (CM), because another way of viewing the intended critique in this central section of the book is twofold: as an "internal" evaluation on one hand and an "external" assessment on the other. The task before us requires the balanced perspectives of both an economist and an ecologist—the biophysical systems that supply the energy for the economic system and the economic principles that use and distribute that energy in the most equitable and sustainable manner possible.

By an internal evaluation, we mean efforts to discuss the structure of economic theory. The central question becomes "how does the practice of economics, as a discipline, lead to policies, actions, and perspectives that result in environmental destruction and resource depletion, both of which are counter to the goals and biophysical principles of sustainability?" An obvious example might be the orientation of conventional methodology to economic growth as an unquestioned and positive goal—something that is embedded in certain branches of economic theory and analysis, regardless of how that principle is employed in practice.

An external evaluation, on the other hand, pertains to the evolution of economic society and the creation of a dominant materialistic and acquisitive culture. Here, economic methodology plays only a background role. To be sure, virtually all real-world economic initiatives, projects, and in fact the mindsets of any community, region, or nation owe their genesis to actual economic principles drawn from the discipline, which should not remain blameless. In other words, economic practices are often used to interpret economic principles to achieve a desired but nonsustainable outcome. Nonetheless, problems arising in this arena stem mainly from human values, attitudes, and behaviors rather than from economic theory.

Of course, there can be two situations wherein we observe destructive or inhumane actions within economic society. First, the philosophy supporting—even propelling—such efforts can *honestly* purport to follow an *incorrect* economic principle. Second, practitioners can *incorrectly* interpret an economic principle, which is not solely to blame. In the first case, economic methodology may be said to be the problem, and in the second, the fault lies more with the misapplication or misinterpretation of economic principles, whether intentional or unintentional. We are concerned with both, because *actions are being taken that threaten socioeconomic sustainability.*

At this point of our admittedly ambitious and far-reaching task, it is useful to restate our purpose and the major arguments that support it. The discipline of economics and its corollary economic practices have been used to create and support a modern world that threatens the very health of the planet and has human society on clearly unsustainable paths. One way of framing this situation is that economics and economic practices have violated the immutable biophysical laws. This section of the book, beginning with Chapter 3, addresses the nature of this violation under the premise that the first step in bringing economics and ecology into harmony is a clear understanding of the problem.

Economics as a field of study began as a necessity. Consequently, it is somewhat misleading to call it a discipline or field of study. These terms carry an academic overtone, which suggests the pursuit of economics might be the choice of an aspiring college student or perhaps a leisure-class intellectual. Moreover, it suggests the categories into which human knowledge has been organized over the centuries, as if taxonomic choices could have been made almost at the whim of some scholar.

The imperative for economics left no such choices. It was a pursuit that addressed human survival. How do humans wrest a living from a bountiful but occasionally begrudging planet? How do people and their collective communities insure their physical survival through good times and bad, vagaries of seasonal change, natural disasters, and the extremes of weather and climatic change? In other words, economics stemmed from the imperative to survive, which is a precondition to everything else—nothing more, nothing less. As such, it was a means to an end, not philosophy, religion, or the arts, despite the rather unfortunate pedestal to which economics and economic life have been elevated in modern times.

Scarcity and Human Survival

The generic term assigned to the situation faced by human populations in this quest to perpetuate their existence was *scarcity*. Scarcity, as a term in economic jargon, means the wherewithal to support life is not superabundant. In other words, the resources, assets, or whatever they might be called require time and effort on the part of people to extract a "living." If a large part of human effort is required to survive, but just barely, that person or those people (family, tribe, community, or nation) may be considered near the subsistence level of income or, in more modern terms, to be *poor*.

On the other hand, a person is clearly better off materially if little of their time and effort is required to survive. In the earliest context, such a person was said to have had much leisure. Only in more modern times have we come to refer to them as *rich*. This is a telling point as we trace the evolution of economic thought from cave dwellers to the present. In an ancient society, the range of goods to consume was obviously limited, and leisure, or free time, is assumed to have been the goal. As modern industrial societies have dramatically increased the range of produced artifacts available, the manifestation of wealth that exceeds subsistence has evolved into the ownership of a wide range of goods—many of which are generally termed *luxury* items clearly unnecessary for survival.

At this point, we will take brief journey through history to help you, the reader, better understand the origin of these dynamics. Evidence indicates that hunting-gathering peoples lived surprisingly well together, despite the lack of a rigid social structure, solving their problems among themselves, largely without courts and without a particular propensity for violence. They also demonstrated a remarkable ability to thrive for long periods, sometimes thousands of years, in harmony with their environment. They were environmentally and socially harmonious and thus sustainable because they were egalitarian, and they were egalitarian because they were socially and environmentally harmonious. They intuitively understood the reciprocal, indissoluble connection between their social life and the sustainability of their inseparable environment.

The basic social unit of most hunting-gathering peoples, based on studies of contemporary hunter-gatherer societies, was the band, a small-scale nomadic group of 15 to 30 people who were related through kinship. These bands were relatively egalitarian in that leadership was rather informal and subject to the constraints of popular opinion. Leadership tended to be by example instead of arbitrary order or decree because a leader could persuade but not command. This form of leadership allowed for a degree of freedom unknown in more hierarchical societies, but at the same time put hunter-gatherers at a distinct disadvantage when they finally encountered centrally organized colonial authorities.[1]

Another characteristic associated with mobility was the habit of hunter-gatherers to concentrate and disperse, which appears to represent the interplay of ecological necessity and social possibility. Rather than live in uniformly sized assemblages throughout the year, they tended to disperse into small groups, the aforementioned 15 to 30 people, that spent part of the year foraging, only to gather again into much larger aggregates of 100 to 200 people at other times of the year, where the supply of food, say an abundance of fish, made such a gathering possible.[2]

Hunter-gatherers were by nature and necessity nomadic—a traditional form of wandering as a way of life wherein people move their encampment several times a year as they either searched for food or followed the known seasonal order of their food supply. "Home" was the journey in that belonging, dwelling, and livelihood were all components thereof. Home, in this sense, was en route.

The nomadic way of life was essentially a response to prevailing circumstances, as opposed to a matter of conviction. Nevertheless, a nomadic journey is in many ways a more flexible and adaptive response to life than is living in a settlement.

This element of mobility was also an important component of their politics, because they "voted with their feet" by moving away from an unpopular leader rather than submitting to that person's rule. Further, such mobility was a means of settling conflicts, something that proved increasingly difficult as people became more sedentary.

Nomads were in many ways more in harmony with the environment than a sedentary culture, because the rigors and uncertainties of a wandering lifestyle controlled, in part, the size of the overall human population while allowing little technological development. In this sense, wandering groups of people tended to be small, versatile, and mobile.

Although a nomadic people may in some cases have altered a spring of water for their use, dug a well, or hid an ostrich egg filled with water for emergencies, they were largely controlled by when and where they found water. Put differently, water brought nomads to it. On the other hand, the human wastes were simply left to recycle in the environment as a reinvestment of biological capital each time the people moved on.

In addition, nomads, who carried their possessions with them as they moved about, introduced little technology of lasting consequence into the landscape, other than fire and the eventual extinction of some species of prey. Even though they may, in the short-term, have depleted populations of local game animals or seasonal plants, they gave the land a chance to heal and replenish itself between seasons of use. Finally, the sense of place for a nomadic people was likely associated with a familiar circuit dictated by the whereabouts of seasonal foods, and later pastures for their herds. Domestication of animals was arguably the beginning of *surplus*.

Although hunter-gatherers had the right of personal ownership, it applied only to mobile property, that which they could carry with them, such as their hunting knives or gathering baskets. On the other hand, things they could not carry with them, such as land, were to be shared equally through rights of use but could not be personally controlled to the exclusion of others or abused to the detriment of future generations.

In traditional systems of common property, the land is held in a kinship-based collective, while individuals owned movable property. Rules of reciprocal accesses made it possible for an individual to satisfy life's necessities by drawing on the resources of several territories, such as the shared rights among the indigenous Cherokee peoples of eastern North America.

In the traditional Cherokee economic system, both the land and its abundance would be shared among clans. One clan could gather, another could camp, and yet a third could hunt on the same land. There was a fluid right of common usage rather than a rigid individual right to private property. The value was thus placed on sharing and reciprocity, on the widest distribution of wealth, and on limiting the inequalities within the economic system.[3]

Sharing was the core value of social interaction among hunter-gatherers, with a strong emphasis on the importance of generalized reciprocity—the unconditional giving of something without any expectation of immediate return. The combination of generalized reciprocity and an absence of private ownership of land has led many anthropologists to consider the hunter-gatherer way of life as a "primitive communism," in the true sense of "communism."

Hunter-gatherer peoples lived with few material possessions for hundreds of thousands of years and enjoyed lives that were in many ways richer, freer, and more fulfilling than ours. These peoples so structured their lives that they wanted little, needed little, and found what they required at their disposal in their immediate surroundings. They were comfortable precisely because they achieved a balance between necessities and wants, by being satisfied with little. There are, after all, two ways to wealth—working harder or wanting less.

The !Kung Bushmen of southern Africa, for example, spent only 12 to 19 hours a week getting food because their work was social and cooperative, which means they obtained their particular food items with the least possible expenditure of energy. Thus, they had abundant time for eating, drinking, playing, and general socializing. In addition, young people were not expected to work until well into their 20s, and no one was expected to work after age 40 or so.

Hunter-gatherers also had much personal freedom. Among the !Kung Bushmen and the Hadza of Tanzania, there were either no leaders or only temporary leaders with severely limited authority. These societies had personal equality in that *everyone* belonged to the same social class *and* had gender equality. Their technologies and social systems, including their economies of having enough or a sense of "enoughness," allowed them to live sustainably for tens of thousands of years. One of the reasons they were sustainable is that they made no connection between what an individual produced and their economic security, so acquisition of things to ensure personal survival and material comfort was not an issue.[4]

With the advent of herding, agriculture, and progressive settlement, however, humanity created the concept of "wilderness," and so the distinctions between "tame" (meaning *controlled*) and "wild" (meaning *uncontrolled*) plants and animals began to emerge in the human psyche. Along with the notion of tame and wild plants and animals came the perceived need to not only control space but also to own it through boundaries in the form of landscape markers, pastures, fields, and villages. In this way, the uncontrolled land or wilderness of the hunter-gatherers came to be viewed in the minds of settled folk either as free for the taking or as a threat to their existence.

So it was that the dawn of agriculture, which arose in the Fertile Crescent of the Middle East, ushered in a new era of controlling land through often-contested boundaries based on a sense of personal ownership. The Fertile Crescent is a crescent-shaped valley stretching from just south of modern-day Jerusalem, northward along the Mediterranean coast to present-day Syria, eastward through present-day Iraq, and then southward along the Tigris and Euphrates rivers to the Persian Gulf. Although sparsely inhabited for centuries, it is thought that agriculture originated in this valley around 8000 BCE. The region was not only greener in those days, but it was also home to a great diversity of annual plants, including grasses with large seeds, such as wild wheat and barley, which grew in abundance. In fact, the wild varieties of wheat have higher nutritional values than domestic wheat.[5] This combination of factors allowed tribes of nomadic hunters, gatherers, and herders to settle along the lush banks of the rivers, where the fertile soil and plentiful water made it possible for them to become the world's first farmers. The rivers also provided fish that were used both as food and fertilizer, as well as giant reeds and clay for building materials.

"One of the most important developments in the existence of human society was the successful shift from a subsistence economy based on foraging to one primarily based on food production derived from cultivated plants and domesticated animals."[2] Being able to grow one's own food was a substantial hedge against hunger and thus proved to be the impetus for settlement that, in turn, became the foundation of civilization. Farming gave rise to social planning as once-nomadic tribes settled down and joined cooperative forces. Irrigation arose in response to the need of supporting growing populations—and so the discipline of agriculture was born.[6]

Around 5000 BCE, the first cities were constructed in the southern part of this long valley, near the Persian Gulf, by an intelligent, resourceful, and energetic people who became known as the Sumerians. The Sumerians gradually extended their civilization northward over the decades to becoming the first great empire—Mesopotamia, the name given to this geographical area by the ancient Greeks, meaning "land between two rivers."[7]

As the farming population grew, groups of people migrated northwestward out of the Fertile Crescent and colonized much of what is Europe today. As they did so, they replaced the indigenous hunter-gatherers, some of which may have taken up farming rather than surrender their home territories to the newcomers. Nevertheless, data indicate that the newly arrived farmers bred at a rate sufficient to keep their population expanding northwestward.[8]

The shift from a hunter-gatherer way of life to one of increased sedentism (the term archaeologists use to describe the process of settling down) and its concomitant social interaction and the maintenance of permanent agricultural fields and irrigation canals occurred in just a few independent centers around the world. One center was the circumscribed upper-middle Zaña Valley of Peru (as opposed to a low, broad valley), where four canals, drawn from hydraulically manageable small, lateral streams, were found on the southern side of the river flowing through the valley. The canals, stacked on top of one another, were dated from 1,190 years ago (the most recent) to about

6,700 years ago (the oldest). Remnants of at least 51 sites of human habitation dotted the intermontane countryside around the canals. Thus began the incipient production of food in an artificially created, wet agro-ecosystem.

Evidence indicates this early irrigation farming was accomplished through communally organized labor to construct and maintain the canals, which necessitated the scheduling of daily activities beyond individual households. Nevertheless, to support the inevitable increase in the local population required an economy wherein farming was combined with hunting and gathering. The commitment to agriculture was more than simply the transition to a sedentary life structured around sustainable, small-scale production of food, it was also the commitment to a set of decisions and responses that resulted in fundamental, organizational changes in society, increased risks and uncertainties, and shifts in social roles as a result of the dependence on irrigation technology.[9]

As indicated by the necessity to schedule daily activities beyond individual households, agriculture brought with it both a sedentary way of life and a permanent change in the flow of living. Whereas the daily life of a hunter-gatherer was a seamless whole, a farmer's life became divided into *home* (rest) and *field* (work). While a hunter-gatherer had intrinsic value as a human being with respect to the community, a farmer's sense of self-worth became extrinsic, both personally and with respect to the community as symbolized by, and permanently attached to, *productivity*—a measure based primarily on how hard a farmer worked and thus the quantity of good or services the farmer produced. In addition, the sedentary life of a farmer changed the notion of *property*.

On the other hand, the growing agricultural lifestyle caused many people to suffer ill health, as illustrated by the analyses of human skeletons excavated at a variety of prehistoric farming villages. As fecal waste from the villagers accumulated, disease and parasites flourished, contaminating water supplies whereby they infected the residents. In addition, the people's skeletal structure became weaker due to poorer nutrition than people experienced prior to the agricultural way of life. Evidence also indicates that infants and young children perished more frequently than they had at the height of the Stone Age.[10]

So, the dawn of agriculture, which ultimately gave birth to civilizations, created another powerful, albeit unconscious, bias in the human psyche. For the first time, humans saw themselves as clearly distinct—in their reasoning at least—from and superior to the rest of nature. They began to consider themselves as masters of, rather than members of, nature's community of life. It seems that farmers had a mindset of utility that was opposed to biodiversity from the beginning—an attitude that still prevails among the world's farmers of today. In fact, wild nature, humankind's millennial life-support system, suddenly came to be seen as a fierce competitor—a perpetual enemy to be vanquished when possible and subjugated when not.[11]

Accordingly, to those who lived a progressively sedentary life as farmers, land became a commodity to be bought, owned, and sold. Thus, when hunter-gatherer cultures, such as the American Indians, "sold" their land to the invaders (in this case, Europeans), they were really selling the right to use their land, not to own it outright as fixed property to the exclusion of others, something the Europeans did not understand. The European's difficulty in comprehending the difference probably arose because once a sedentary lifestyle is embraced, it is almost impossible to return to a nomadic way of life, including the thinking that accompanies it.

Until fairly recently, historically speaking, property in Britannia, as early England was known, used to be a matter of possessing the right to use land and its resources, and most areas had some kind of shared rights. Today, the land is considered to be property, and the words for the British shared rights of old have all but disappeared: "estovers" (the *right* to collect firewood), "pannage" (the *right* to put one's pigs in the woods), "turbary" (the *right* to cut turf), and "peccary" (the commoner's *right* to catch fish) are no longer in the British vocabulary. Now, while the landowner's rights are almost absolute, the common people no longer have the right of access to most lands in England.[12]

Even the future of agriculture in the Fertile Crescent is becoming increasingly grim due to a combination of a changing global climate and continual diversions of the water far upstream in both the Tigris and Euphrates rivers. In 2009, after 2 years of drought, which experts warn could become permanent, farmers began abandoning their fields, even as the Turks began tightening the grip of their dams, which have already reduced the rivers to mere trickles. In addition, however, new Iranian dams are further reducing the flows in these two historic rivers.[13]

We, as individuals, may despair when we contemplate the failure of so many earlier human societies to recognize their pending environmental problems, as well as their failure to resolve them—especially when we see our local, national, and global society committing the same kinds of mistakes on an even larger scale and faster time track. But the current environmental crisis is much more complex than earlier ones, because modern society is qualitatively different than previous kinds of human communities. Old problems are occurring in new contexts, and new problems are being created, both as short-term solutions to old problems and as fundamentally new concepts. Pollution of the world's oceans, depletion of the ozone layer, production of enormous numbers and amounts of untested chemical compounds that find their way into the environment, and the potential human exacerbation of global climate change were simply not issues in olden times.[14] But they are the issues of today.

Although a few cultures (such as Bedouin clans in the Middle-Eastern deserts and the Lapland reindeer herders) still live lightly on the land, most of humanity leaves a heavy footprint, consuming nearly a quarter of the Earth's biophysical productivity. In fact, land use continually transforms Earth's terrestrial surface, thereby resulting in changes within biogeochemical cycles and thus changes in the ability of ecosystems to deliver services critical to human well-being.[15]

Thus, while the hunter-gatherers created *the right to use common property* spontaneously in their living (the "commons," which is considered to be everyone's birthright), it is today being progressively eroded as people, especially in the industrialized countries, are evolving from *Homo sapiens* (modern human) into *Homo economis* (economic human). To arrest this erosion, we must understand and accept that the quality of our individual lives depends on the collective outcome of our personal thoughts, decisions, and actions as they coalesce in the environment over time, particularly with respect to the global commons.

Thinking about the evolutionary process surrounding the notion of surplus yields vital points in understanding modern economic life and the economic methodology that supports it. This task is undertaken more fully in the following chapter, which deals with the notion of consumption.

Economics and Human Nature

The first task of any academic discipline or field of study is to identify its working theory of human nature. What does it mean to be "human"? The arts emphasize the creative urge, the humanities our philosophical and reflective bent. Biology will note that we humans are one among many living organisms, and so on. Economics is considered to be a social science, which means the issues and topics that compose the core of the discipline innately involve human behavior. Political science emphasizes the ways in which humans organize and operate in groups to achieve power and influence. Sociology does much the same for social or communal behavior of all kinds, and anthropology is more or less sociology and culture throughout history. Psychologists, who often have problems determining whether the discipline is a natural or a social science, explore human motivations and internal, mental processes. Each of these fields of study, as branches of the social sciences, tends to adopt and refine a theory of human nature that fits the purposes of the discipline and can be termed *self-serving* in that sense.

It is no different with economics. Given the perceived need for mathematical certainty and predictability, the profession of economics has responded with a bare-bones, almost mechanical image of a human being. Other social sciences, and certainly the arts and humanities, honor and even celebrate the intuitive, creative, capricious, and often emotional and contradictory nature of the human psyche. Emotional variation is the stuff of a fully developed life. Economic theory, however, cannot accept uncertainty, which heralds unpredictability. The result is a construct that has become known as "Rational Economic Man." It is useful to examine some historical background that assists in detailing that image.

Rational Economic Man

For better or worse, the beginnings of economics as a discipline supposedly began with Adam Smith (June 5, 1723 to July 17, 1790) and publication of his *Wealth of Nations* in 1776. Staunch devotees of a market economy often note, with patriotic fervor, that the birth of economics is coincident with the beginnings of the United States, which was destined to provide the most aggressive testing ground for these classically liberal ideas. Attention to economic life, of course, goes back much further.

It is significant that Adam Smith was every inch a product of the Enlightenment, which, over a period of three to five centuries, ushered in an extended love affair with the notion that human analytical thought can, among other things, uncover absolute truth—something that is impossible through knowledge because knowledge is always relative and perpetually outdated. This culminates a process marked by shifts in the dominant view of Western society from reliance on faith to reliance on reason, from divine revelation to empirical scientific discovery, and from the concentration of human affairs within the realm of the transcendental and spiritual to the realm of the secular and material.

As economic theory began to deviate from the Middle-Age doctrine of the Divine Right of Kings, Western society experimented with the notion that it could discover physical principles of the world, not simply by decree but through observation and reason. The people who were experimenting with economic theory also held the notion that they could structure their own economic and political institutions to be responsive to the democratically expressed will of the people, and that ruling institutions could be made to serve the people, rather than the reverse.

Thousands of volumes have been written on all aspects of these powerful ideas, and it is not our purpose to recreate the major tenets of intellectual history. The important point is that, as a typical product of the times, economic methodology became enamored from its outset with empirical scientific method and thus preoccupied with the self-appointed task of becoming a science.

Social philosophers of the 18th-century Enlightenment were so bedazzled by the astounding scientific discoveries of the 17th century in astronomy, physics, mathematics, plate tectonic geology, chemistry, and so on, that many set off in a search for what was termed the universal "laws of motion" of human behavior (the immutable "nature" of humans, if you will). Certainly, the reasoning went, there must be deterministic laws for human behavior that parallel Isaac Newton's law of gravity. As a species, we have always harbored a fascination in playing with a novel, new toy.

And the chosen view of science is part of the problem. Although humankind, having spread throughout the globe, is perhaps the most adaptable species in the history of planet Earth, that very success is proving to be a systemic failure, because we, in Western industrial society, are today imprisoned in the hubris of Baconian tradition. Francis Bacon (January 22, 1561 to April 9, 1626), considered to be the father of modern scientific inquiry,[16] viewed nature as a "common harlot" and thus urged future generations to "tame, squeeze, mold," and "shape" nature in order to be "the undisputed sovereign of the physical world." This view gave rise to the reductionist lens through which we modern humans intellectually dissect the world into isolated, competing fragments rather than risk our deeply human emotional-spiritual engagement with the dynamic, life-support systems that sustain us.

If, therefore, this new body of analytical methods was to be considered a science, it must somehow be able to ascribe absolute predictability with respect to the material, acquisitive actions of human beings. Thus began the reductionist lens of economics—the defining moment when yesteryear's material success foreshadowed today's systemic failure. After all, an apple *always* falls from a tree in the same manner. The resulting bundle of assumptions, known as Rational Economic Man, has the following major characteristics:

- Is self-interested (self-centered)
- Has complete and perfect knowledge of all consumption and production alternatives
- Is acquisitive
- Is materialistic
- Always prefers more to less of any economic good or service
- Prefers immediacy—something now is always preferable to the same thing later
- Always makes the same choices ("rational")
- Is competitive in behavior, both in consumption and production (employment)

Refinement and exemplification of these features will occur in the next two chapters, as we separately discuss consumption and production. Further, those features identified above will be particularly useful in exploring the full implications of the preoccupation of our society with the phenomenon of economic growth as part of its philosophical underpinnings. Suffice it to say, Rational Economic Man stands as the definitive image of human nature emanating from the classical beginnings of the discipline of economics. Squeezing out love, variability, and altruism or charity results in the image of *Man as Machine*, pure and simple. As we will see, this has had profound effects on the evolution of economic thought throughout the years and, consequently, on the society that it seeks to serve.

From Necessities to Wants and Subsistence to Wealth

An organism driven by aforementioned characteristics can scarcely be content with a subsistence life. An acquisitive, self-interested, competitive, materialist is hard-wired to seek more. To an astute observer, this suite of characteristics is pretty much the demise of economics as a tool kit for mere survival. The initial notion of an economy as a functional means whereby to "meet necessities" quickly transitions to a goal of "satisfying wants." In a sense, there is nothing wrong with this. (Of course, the modern theory of demand steadfastly contends that you cannot tell the difference between wants and necessities—nor should you want to.) The creation of leisure due to the fact that economic needs are met more easily than was possible in the past is a badge of success for economic activity. It has supposedly done its job. As an aside, all priceless treasures of the past were created because people and their societies carved out leisure time over and above the efforts needed just to insure their survival. The Parthenon and the Sistine Chapel are products of surplus and leisure.

But this set of assumptions, which supposedly define the economic behavior of a human being, includes no logical stopping point. The possibility of ultimate consumer satisfaction is not acknowledged, and limits on acquisition of goods and wealth do not exist. Therefore, we have seen the continuous evolution of a discipline that was initially intended to show people and societies how to subsist into one that purports to show them how to become wealthy. The blueprint for acquiring enough to survive has been transformed into a model for acquiring any and all goods available in the modern world. Necessities are transformed into *wants*. Comfort is replaced by material luxury. And all the while, the internal workings of the discipline offer no assistance in making the essentially philosophical decisions in determining when one has enough. The credo is left at "More Is Better." Consequently, any movement from this morally bleak, unsatisfying goal is left to the strictly noneconomic sphere. In essence, as the lust for material objects grows, the notion of *enough* progressively evaporates.

The lack of acquisitive instincts on the part of the particular actors in the economic drama (businesses and people) would cause the predictive models to malfunction. It is as though the Rational Economic Man paradigm has conditioned people to behave like trained animals in a circus. Provide certain stimuli, and a standard, completely predictable response occurs. The show will go on, and people enjoy not only watching but also participating in the process. However, if a different or unpredictable response follows a standard stimulus, the circus may have to be shut down. It can even be dangerous for the trainer.

At the microeconomic level, therefore, perpetual acquisition of material wealth remains the sanctioned behavior. People are assumed to be almost wholly materialistic and perpetually willing (and eager) to compete in bettering their economic lot. At the macroeconomic level, the system is pushed in directions that will best seem to accommodate the personal drive for more. Government, after all, is said to ideally follow cultural norms, not lead them.

Misuse of Economics in Practice

Thus far, the critique of economics in this section has focused on economic methodology. We have attempted to comment—if not in detail—on the development and structure of economic theory as it tends to support and promote an unsustainable system. Alternatively, this can be seen as attempting to uncover and analyze how the beliefs of economists, acting as the "high priests" of the discipline, might convey and promote a less-than-desirable conventional wisdom.

There is, however, much more to the story. *Economic practice* as it has evolved among real-world decision makers is also greatly to blame for many excessive and ruinous practices. To be sure, if we observe some destructive feature, such as environmental degradation, or a clearly unfortunate effect of overconsumption, it is a fine and indistinct line between ascribing the ultimate blame to the structure of conventional economic wisdom or to the excesses of some element of economic society.

On one hand, it is fashionable to ignore the ivory tower priests. Business and (even) government leaders, charged with real-world decision-making responsibilities for allocating resources in one direction or another, love to point out the difference between theory and practice: "That's fine in theory, but that's not the way it works in the real world," or, "Have they ever met a payroll?" This is simply saying that what textbooks expound is one thing, but when faced with decisions that affect people, jobs, incomes, necessary services, and so on, different, hard-nosed criteria must be employed. We do not have the luxury of theoretical elegance.

On the other hand, and we want to be crystal clear on this point, even though the *direct* influence on a particular real-world attitude or decision may be invisible, the *indirect* effects can often be traced to some hallowed economic principle. Of course, economists are accustomed to assuming their discipline is a complete omnibus system wherein they have, in effect, thought of everything. Thus, when an undesirable or unfortunate feature of modern life is pointed out, they are likely to contend that the problem stems from an incorrect application of economics rather than from the actual core teachings, which should be held blameless. Yet, it is also true that those promoting the particular personal or institutional behavior creating the negative effects (i.e., those creating the problem) almost invariably ascribe their actions to some standard economic principles. *Is it theory or practice that must assume ultimate responsibility?* Let us further explore this game of finger-pointing and denial through three examples.

EXAMPLE 3.1

Suppose we observe a serious issue of water pollution emanating from a manufacturing plant in some city, which in turn has impaired recreation, fishing, wildlife, and even water supply for downstream communities. Imagine a bare-bones conversation between an environmentalist and an economist:

Environmentalist: Your discipline blindly recommends producing more goods and services as the key to social welfare, no matter what the consequences. Now look at the mess we have as a result of so-called "unintended consequences" of these private-production activities.

Economist: No, we recommend that in a system of private ownership and production, all costs of production must be borne by the producer, and then incorporated into price of the product. What we have here is a failure to do that. Degradation of the water quality is clearly a cost, and it is not being shouldered by the firm but passed on to other innocent bystanders. Our theory recommends internalization of all costs; so the problem is really the failure to follow our theory, not the theory itself.

Environmentalist: But your theory calls these types of things externalities, as if they're an unimportant side issue to the main business of producing and consuming. In effect, you're urging society to ignore or minimize their impact.

Economist: But we give clear guidelines for internalizing these externalities. If this is not done, then they clearly are overproducing that product, since the price to consumers is artificially low because it does not include all 'production' costs. If internalization were to occur, the price would accurately reflect these costs (and also profits would be lower), and the result would be socially optimal.

Environmentalist: Given the quantitative measurement fetish of your discipline, it's fine to say loftily that all costs, such as air or water pollution, should be incorporated, but those are difficult or impossible to price accurately. Especially when compared to profits, jobs, prices for products, and similar "hard" economic quantities, the intangible environmental factors, which erode quality of life for everyone, tend to be minimized or even forgotten completely.

Economist: Give us time. We're working on better measurement techniques.

This hypothetical exchange has taken place, in one form or another, in virtually every community. The root cause of some problem is ascribed to economics, and the discipline is defended through a contention that the real problem is the inappropriate application of the theory, and not the theory, which may even be defended as the solution, if only it were followed correctly. The self-image of the economist is even probably as an environmentalist.

EXAMPLE 3.2

Suppose that a concerned mother is arguing with an economist about whether strict self-interest always dictates her every action.

Concerned Mother: I'm sacrificing now and giving up things I want so that my kid can go to college and make something of himself. That restricts my options now for his sake. That's not self-interest. It's more a form of altruism or selflessness. Your theory doesn't allow that.

Economist: Of course that's self-interest! Your family is just an extension of yourself, and you would derive great pride and satisfaction from seeing your son go to college. Besides, if he gets a good education, and then a good job, you won't have to support him in the future, and you'll have more later. He may even support you in your old age. Your transfer of money to your son isn't a market transaction, but it is thinly disguised self-interest behavior.

Concerned Mother: I'm not sure I buy that, but let's try this: I'm concerned about those poorer than I, and I routinely donate money to the local homeless shelter. Certainly, you would agree that can be termed altruistic behavior, which you economists pretend doesn't exist.

Economist: Well, not exactly. Of course, you're just making moderate donations of money you might not even miss. You're not giving them your entire income. And I'm sure you get some satisfaction out of feeling 'charitable'—maybe even equal to the loss of income. Further, by keeping these people off the street, you minimize other costs, such as reducing the likelihood of getting robbed. And of course, you get a tax deduction.

Concerned Mother: You guys are heartless! Given my complete disgust with this selfish dog-eat-dog consumerist society, what if I just drop out, grow my own food and refuse to play your silly materialistic games?

Economist: Not a problem. You may appear to be an anomaly, but it won't matter. Everyone else will still continue to consume and compete. My models will still work fine for the economy—that's just the way people are. You're free to do your thing.

This exchange questions the underpinning assumptions about human nature that support standard economic theory. The economist would contend that although apparently anomalous behavior might appear to occur in the short run (e.g., a bit of income to charity), it would not hold for all ranges of income and would not persist in the long run. Also, as long as others display the predicted behavior, the models work accurately and thus describe society. Further, the economist would defend the methodology by construing any apparently altruistic or anomalous behavior simply as disguised self-interest.

EXAMPLE 3.3

Suppose a situation exists wherein a community activist interested in promoting a sustainable and healthy local economy is complaining to an economist about the effects of globalization and the resultant ruinous outsourcing of jobs to China or Mexico.

Community Activist: We seem to import everything we consume these days. It's killing us not to produce in this country more of the goods that we consume.

Economist: That's not true! David Ricardo's theory of comparative advantage is one of the oldest tried and true principles in economics. Specialize in what you do best and then exchange with other countries or regions for what you need. Everyone benefits from trade—consumers and productive workers alike.

Community Activist: But we have no jobs left. We're being asked to consume imported goods, and we have little purchasing power with which to do that. Doesn't the theory say we're supposed to export a bunch of stuff to China to balance it out? The data show we're importing at least 10 times as much as we're exporting.

Economist: Well, that may be true for now, but that will change as the Chinese consumer culture develops and they want more of our stuff. At present, though, we should keep the volume of trade at high levels, and concentrate on 'knowledge-based' jobs that we do best, not on manufacturing.

Community Activist: But what's happening is not trade. Our corporations are relocating there for the cheap labor, then just bringing stuff back here to sell. It can't work in the long run. There's something wrong with your notion that, "Trade helps everyone."

In this example, a basic tenet of economic theory is questioned. No principle is more fundamental to market-based economics than the notion of *free* and voluntary exchange. (Here, it must be noted that *nothing* in the entire universe is *free*, because everything is constrained by its relationship to everything else—including every component of every economic system, despite the contention of economic theory to the contrary.) This tenet, which can only be construed as *voluntary*, underlies the entire theory of supply and demand. The widely accepted (and over 200-year-old) theory of comparative advantage simply extends that principle to trade between nations—in addition to the familiar market exchanges between buyers and sellers.

In recent years, however, the phenomenon of globalization, and the rising influence of transnational corporations, may have fundamentally changed the reality. These corporations, by being involved in both the producing and consuming countries, are able to keep—as profit—the incremental surplus that once accrued to the workers through higher wages and to the consumers through lower prices.

> Without delving further into this intriguing hot-button issue, our point here is that due to the evolution of new institutions and their completely predictable profit-maximizing behavior, the theory that once accurately described a simpler and earlier world may no longer hold. Very few "immutable principles" based on human behavior remain immutable forever.

Our task, thus far, has been to identify different ways in which economic theory or economic practice can lead societies astray and foster an unsustainable system. We have uncovered, through these hypothetical examples, different types of cases wherein undesirable effects on our world (local, national, and international) can realistically be ascribed to economics per se, or to economic behavior of one type or another. If the accuracy of the theory is the problem, we may conclude that prescribed changes to economic doctrine are in order. If the problem is either a societal one or an incorrect interpretation of the theory, the behavior of people and institutions must change. The categories can be summarized as follows:

- Incorrect or counterproductive economic theory
- Incorrect or exaggerated use of the theory
- Due to economic evolution or institutional changes, an out-of-date theory
- Underlying assumptions that cannot be granted

Most observed problems of an economically related nature probably involve elements of more than one of the above categories. The underpinnings for the intellectual rationalization of the cause of a damaging action or project can be based on bad behavior, poor or out-of-date theory, and false assumptions—all at the same time. As we explore the innate nature of economics and some of the ways it can mislead, we must be on alert for such fallacies. But countervailing defenses will arise promptly, due to the perceived threats to one's ego and the high monetary stakes. After all, as the theory maintains, self-interest and self-preservation are powerful forces.

Growth as Economic Religion

Finally, as a result of all this, and in large part actually because of the flaws in economic theory or in overly competitive or self-interested economic behavior, the cultural and socioeconomic bottom line has become the institutionalization of the "growth ethic." This becomes the norm, despite the environmental impacts on the resource base; community land use, such as the location of transportation corridors; on the inequalities of wealth and income; or on relations with other nations. Fostering uninterrupted growth has become the effective public religion. No matter what the economic question or the problem might be, continued growth is extended as the solution.

Perpetual growth has become a societal requirement for both economic theory and practice in several ways. First, the individual imperative for Rational Economic Man demands it. Individuals and the communities within which they operate value a culture of continual accumulation. Second, businesses and the producing sector seek to produce ever more goods and services, and to grow in size as a result. Ever-increasing business profit is revered as a good and necessary component of all this.

Observe the current joint crises we noted in establishing the premises of this book. All one hears in the media is the necessity of "getting this country growing again" and "putting people back to work." This is an appealing siren song, because people clearly want to work, and they clearly want to consume. There is the occasional mention of *climate change*, but in the currently charged political atmosphere, it is often just a question of whether it is even real or not.

It is our firm opinion that current climate change is demonstrably real, as evidenced by the global melting of glaciers and arctic sea ice, and that we, as a society, are destined for deep disappointment if we rely on the traditional growth ethic to restore prosperity. Economics as religion is a barren and tragically misleading philosophy.

There are many directions the analysis could go from here. Beginning from the basic nature of economics as a discipline, we have raised questions about equity and fairness, environmental effects, national economic goals, and foreign policy—virtually all of the most important questions facing our society. The majority of these threads will be traced more fully in later chapters, but the role of this chapter has been to set the stage by identifying the core of economics as a discipline: what is assumed, how it is employed, and what it can logically be expected to cause through its use. In order to pursue these ideas in more depth, we turn now to focus on the notion of *consumption*.

Endnotes

1. John Gowdy. Introduction. In: *Limited Wants, Unlimited Means,* John Gowdy (ed.), xv–xxix. Washington, DC: Island Press, 1998; Marshall Sahlins. The Original Affluent Society. In: *Limited Wants, Unlimited Means,* John Gowdy (ed.), 5–41. Washington, DC: Island Press, 1998.
2. The foregoing two paragraphs are based on: Rebecca Adamson. People who are Indigenous to the Earth. *YES! A Journal of Positive Futures.* Winter (1997):26–27.
3. John Gowdy. Introduction. In: *Limited Wants, Unlimited Means,* John Gowdy (ed.), xv–xxix. Washington, DC: Island Press, 1998.

4. Cristobal Uauy, Assaf Distelfeld, Tzion Fahima, and others. A NAC Gene Regulating Senescence Improves Grain Protein, Zinc, and Iron Content in Wheat. *Science* 314 (2006):1298–1301.
5. Tom D. Dillehay, Herbert H. Eling, Jr., and Jack Rossen. Preceramic Irrigation Canals in the Peruvian Andes. *Proceedings of the National Academy of Sciences* 102 (2005):17241–17244.
6. Stacey Y. Abrams. The Land between Two Rivers: The Astronomy of Ancient Mesopotamia. *The Electronic Journal of the Astronomical Society of the Atlantic* 3 (no 2) (1991). [No page numbers given.]; The Fertile Crescent. http://visav.phys.uvic.ca/~babul/AstroCourses/P303/mesopotamia.html (accessed October 25, 2010).
7. *Ibid.*
8. Wolfgang Haak, Peter Forster, Barbara Bramanti, and others. Ancient DNA from the First European Farmers in 7500-Year-Old Neolithic Sites. *Science* 310 (2005):1016–1018.
9. The preceding two paragraphs are based on: Tom D. Dillehay, Herbert H. Eling, Jr., and Jack Rossen. Preceramic Irrigation Canals in the Peruvian Andes. *Proceedings of the National Academy of Sciences* 102 (2005):17241–17244.
10. Bruce Bower. Evolution's Ear. *Science News* 174 (2008):22–25.
    Wolfgang Haber. Energy, Food, and Land—The Ecological Traps of Humankind. *Environmental Science and Pollution Research* 14 (2007):359–365.
11. *Ibid.*
12. George Monbiot. Land Reform in Britain. *Resurgence* 181 (1997):4–8.
13. Steve Newman. Fertile Crescent Decline. Earthweek: A Diary of the Planet. *Albany Democrat-Herald, Corvallis (OR) Gazette-Times*. August 14, 2009.
14. Gus diZerega. Re-thinking the Obvious: Modernity and Living Respectfully with Nature. *Trumpeter* 14 (1997):184–193.
15. Helmut Haberl, K. Heinz Erb, Fridolin Krausmann, and others. Quantifying and Mapping the Human Appropriation of Net Primary Production in Earth's Terrestrial Ecosystems. *Proceedings of the National Academy of Sciences* 104 (2007):12942–12947.
16. Francis Bacon. http://en.wikipedia.org/wiki/Francis_Bacon (accessed on November 3, 2010).

4

Consumption Theory

Although the primary meaning of the term *consumption* is to eat or drink, the secondary connotation is to "use," which is how the term is employed in economics. With this in mind, Chapter 4 is composed of four sections: (1) consumption for survival, (2) consumption in practice, (3) affluence as an unmitigated public good, and (4) toward an economics of enough.

Consumption for Survival

Even the word *consumption* is a misnomer. As the laws of thermodynamics indicate, nothing is finally consumed—only transformed. This may seem like a point of picky semantics—an example of the unimportant debates over terms in which ivory tower intellectuals love to engage. But it is not. In fact, it is indicative of what is perhaps the major disastrous error that we have made in structuring society along the lines indicated by the discipline of economics. Let us explain.

The purpose of economic activity is to meet the material needs of human beings. Originally, the economic problem was mere survival, and the term *consumption* meant to meet the necessities of life. In the development of economic methodology, however, it has become much more in that humans have been reduced to consumers. The act of relegating a human soul, with all his or her emotions, values, aspirations, and creativity to a mere processor of material goods is a travesty in itself—but that is another story.

For our purposes here, we focus on a person's demand-side role in the economic sphere and in dissecting and understanding the workings of the economy, which is as the final *consumer* of goods and services. And how, one might be tempted to ask, is this supposedly a disastrous error? To facilitate addressing this question, we reconstruct one of the most common images employed in the basic exposition of the economic system, the *Throughput Economy*. This model depicts a process wherein one begins with natural resources and moves through to the supposed final consumption of a good—depending on which meaning of the term is appropriate.

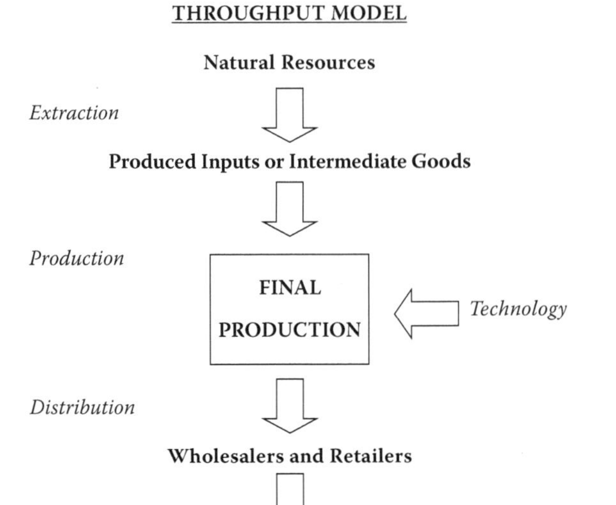

In this model, we begin with the endowment of natural resources available to support the economic activity in question. As we stated earlier, these resources are essential components of the natural environment. The economic process then begins with extraction of the resources (minerals, timber, water, fish, medicinal plants, and so on), the conversion of those resources through the production process ⇒ to intermediate and then finished products ⇒ to the distribution of those goods ⇒ to sales outlets ⇒ and ultimately to the consumer, which may be a household or an organization, such as a business, government, or nonprofit entity. That final product is supposedly then consumed, and in the lexicon of economic methodology, results in value, which can be alternatively expressed with several different terms, such as *satisfaction, well-being, utility,* and *human welfare,* to name a few.

The simple point is as follows: *Natural resources are employed to satisfy human necessities, as well as desires.* Thus, the movement of materials through the economy—from extraction to final consumption—results in the use (consumption) of nature for the pleasure of human beings, however else it may be construed.

In a real sense, of course, this is exactly what the economy does. It employs available resources to meet the material requirements of people. But the story does not end here. In the first place, the implication is that the natural environment holds no value for people until it is somehow "processed" into constructed artifacts. This innately ignores appreciation of nature, wilderness, or biodiversity for its *intrinsic* value. Only the *extrinsic* value (economic conversion potential) of the built or produced environment seems to matter.

But, as destructive as that may be, it is not the main misleading drawback of the throughput model. The major problem with the throughput model is that if the implications of the construct are followed, the materials extracted from nature at the beginning of the conversion process are in effect treated as though they disappear from the face of the Earth with the act of consumption. When anything is consumed, it is gone. Foodstuffs and their packaging vanish. Old automobiles vaporize when they no longer run. And, all traces of impacts at any point in the economic chain disappear. Mine tailings, environmental costs of manufacturing, air pollution, and resource depletion due to multiple requirements for transportation do not exist. In short, and again, ignored is the fact that in the real world, *nothing is finally consumed, but rather everything is merely converted to something else.* In the "perfect" world of the throughput model, there are *no* environmental impacts or resource depletion.

The significance of treating the natural environment in this manner cannot be overstated. The core of economic methodology ignores any environmental effects of extraction, production, distribution, and consumption—the vital activities of carrying on any economy, whether that economy is sophisticated or relatively primitive. Of course, over the last few decades, it has become clear that the environmental effects of extraction, production, and distribution, cannot be ignored—except at our collective, long-term peril. Thus, we have seen the development of such subfields as environmental economics and ecological economics. It is indicative, however, that the methodological treatment of such effects proceeds under the term *externalities.* (This branch of economic theory is afforded separate attention in Chapter 7 and will not be further discussed here.) Suffice it to say, common sense dictates that any field of study that labels as "external" the factors that threaten our way of life and the health of the planet, should not be turned loose to direct the material affairs of humankind.

Consumption in Practice

This section consists of two parts: (1) from necessities to wants and (2) assumed insatiability.

From Necessities to Wants

We may return periodically to other implications of the Throughput Model, but it is necessary to explore further the evolution of the concept of consumption. From the earliest subsistence-oriented notion of consuming for survival, the concept has undergone much expansion. The notion of meeting basic requirements is continually evolving. At first, the expansion of the notion of "necessities" grew simply to "needs" and then to "wants," which are today expressed as "demands." After all, once economic activity is sufficient to insure survival, what harm is there in acquiring a bit more than basic subsistence to add a modicum of pleasure to life? As such, music, the arts, and many other forms of social interactions are universally recognized as desirable and life-enhancing qualities.

It would be impossible to trace the genesis of this seductive notion in the human psyche—namely, the desirability of acquiring or consuming more than just the necessities of basic survival. No doubt, the idea goes back millennia—not just centuries—and it is widely believed that acquisitiveness is an innate human trait. Although we doubt this premise, it is a debate for philosophers and the broader range of social scientists.

Let us be clear. There is no more fundamental point anywhere in this book than the notion of *surplus*. Surplus can be defined as the availability of goods or services over and above those required to just break even. This is true whether breaking even is considered from the point of view of the individual (consumer) or from a business (producer). As we will see, many of the fundamental notions of economics spring from this seemingly simple and innocuous idea.

In a production sense, breaking even is a position of zero profit, or just enough to keep the business in existence. Once the firm creates a surplus in the form of profit, the issue of distribution raises its ugly head. Do we pay it out in dividends to the shareholders (owners), or do we raise the wages of the employees (workers)? Thus, the issue of surplus gives rise to the age-old owner versus worker dispute (Proletariat versus Bourgeoisie to Karl Marx), which is at the core of capitalism.

Let us explore further the economic notion of distribution. Think about it. If everyone in a given defined community (e.g., family, tribe, village, or region) has just enough goods (or income) to meet daily necessities, many issues do not arise. Everyone has enough, but no more. There are, for instance, no rich or poor. Moreover, the distinction does not even arise. Once a surplus exists, however, the questions of who gets what and how it is used—as well as all the potential conflicts over these questions—arises. This is a consumption-based way of viewing the ramifications of surplus.

The use of such surplus has defined civilizations throughout recorded history. For example, the Romans extracted resources, or "riches," from their far-flung empire to support a lavish lifestyle for the ruling elite in Rome. Another famous example of the use of surplus centers on the feudal society during the Middle Ages. The Church, as well as the nobility in England and on the continent funded grand palaces and cathedrals while the peasant class and serfs survived at bare subsistence levels. This class-based arrangement (despite some governmental modernization as a result of the Magna Carta and the Enlightenment) evolved into the upper-class/working-class structure following the Industrial Revolution, which led into Victorian England.

All these social structures were determined by choices made on the use of surplus. The disintegration or outright downfall of any social group can be traced in large part to abuses resulting from extreme concentration of surplus, or wealth, of the ruling class. An obvious example is the objection of Martin Luther to abuses by the Church, which led to the Protestant Revolution. This leads to one reason the American experiment of a broad-based, middle-class society is so uniquely noteworthy in human history: It embodies the premise that surplus should be shared by all, even though it embraces a capitalistic economic system. In summary, the role of surplus throughout the ages cannot be overstated as one seeks to understand either the evolution of economic history or the discipline of economics. We do not purport to solve, or even to discuss in any depth, these issues here. Our focus at this juncture is on the structure of economic methodology. The critical point is that consumption above basic requirements is seen as a desirable goal by economic theory.

Assumed Insatiability

All the models conclude that success is based on the amount of consumption—the higher the consumption, the greater the success. This notion goes back at least a century and a half and can be contributed to the ubiquitous methodological push for quantification, an analytical trait of economic analysis that required something to maximize as a measure of "success." Higher mathematics begins with calculus, and calculus is the mathematics of optimization—or, in other words, maximization and minimization.

Over the years, a subtle transformation has occurred. At first blush, how could one more logically choose something to maximize than human satisfaction or well-being? After all, facilitating survival is the original, stated purpose of economics. But the question arises of where to stop. Put differently, when is enough, enough? Do people reach a point of satiation, at which they conclude: "That's enough." If so, this is a problem of sorts for mathematics. Having some quantity simply maximized by increasing forever does not make sense. Maximization is normally subject to some constraint, and the chosen constraint for economic methodology is income.

Thus, the task of the individual citizen, or worker/consumer, is to maximize the individual's level of satisfaction subject to his or her fixed income. This methodological premise gives rise to a frequently stated version of the economic problem as unlimited wants against limited means. Stated differently, the consumer is assumed to always want more, no matter how much the consumer already has, despite being constrained by his or her fixed income.

What if it is pointed out that, either due to the nature of the product or the existence of a *very high* income, having more gives no satisfaction? At this point, the economic answer becomes adroit: "Well, even though they are satiated with *that* product, there are always other goods and services for which there is positive demand." The individual is therefore *assumed* to be permanently acquisitive, no matter what. Nevertheless, from a systemic point of view, while some individuals are satiated with a particular product, which makes it *noneconomic* for them, there will be many others who not only demand that product but also are willing to pay for it. Thus, the product does not become a free good.

The upshot of all this is that all goods and services retain a positive price (defined by economists as "scarce"), and every person is assumed to be permanently acquisitive and in a perpetual state of dissatisfaction—a state that is continually reinforced by advertising. The economic assumption is that, regardless of their income, no one can ever get enough of all the goods and services.

There are some powerful critiques of economics that follow from these methodological observations. First, it justifies an incentive system wherein people are always seeking better jobs and higher income. Second, competition as an ethic becomes the behavioral norm for both individuals and businesses. Third, people are assumed to behave in certain ways for the convenience of the optimization techniques. Would not the reverse be healthier—with techniques chosen to accommodate an innate human nature? This discussion of consumption raises fundamentally important issues, such as the age-old distinction between competition and cooperation. The implications for the value structure and psychological character of all of us as individuals are huge, and many of the themes of this book flow from these ideas. For the moment, however, we turn briefly to the implications of the macroeconomy.

Affluence as an Unmitigated Public Good

Whatever the situation with the makeup and motivations of individual participants in an economy, or citizens in a society, there is the larger question of public policy. How do we organize collective action in the necessary tasks of managing the commons? What operating assumptions exist for government; what direct the actions of public employees; or what constitutes the campaign rhetoric of candidates for public office? A leader is effective only to the degree the leader's stated operating principles mirror those of the populace to be led, served, or represented. Ergo, the questions become the following: How do the above assumptions affect the *culture*? and How, in turn, does this affect the institutions of that culture?

The answers can be stated in many ways. To wit, an effective leader is in concert with public opinion. On the national scene, this can be said to be consistent with "the national character." For the United States, going back at least as far as the early 18th century writings of Alexis deToqueville, this raises long-standing questions of historic tradition. The search for the compatibility values not only captures the efforts of all seekers of public office but also requires commonly accepted norms for the operation of government.

These observations on the nature of consumption, as treated in economic methodology, clearly have implications for the common acceptance of particular cultural traits, which compose the construction of the national character. First, people are assumed to always want more than they have and experience a perpetual tension to better their position. This includes earning more money, thereby improving their economic lot, and for businesses, producing more and thus earning more. Competition and the pursuit of self-interest are not only taken for granted but are even promoted.

Whereas the maximization of personal acquisitiveness is blithely assumed to be limitless, an individual is encouraged to do everything in his or her power to further his or her self-interest (i.e., get a better job). Therefore, promoting each individual's quest for more counts as acceptable public policy.

All this funnels down to a public-policy emphasis on pure *growth*. Expand the pie so *everyone* can have a larger piece, whatever the effects might be, which means the concentrations of wealth tend to be ignored. In fact, the goal effectively becomes undifferentiated growth of aggregated production, with little attention to the composition of goods and services, the *quality* of that production, or the issue of who gets what. If individuals are assumed to always want more material goods (no matter how much they already have) and to have the freedom to choose whatever they want more of, then the compatible version of a macroeconomic policy is simply to maximize gross domestic product (GDP). In this sense, more GDP is better, and less GDP is worse, no matter what resources are used or what the nature of the goods in question might be.

Simply invoking the term *growth* raises a multitude of important issues—many of which have been debated for decades. A thoughtful, comprehensive treatment of growth, along with its meaning and implications as a logical corollary of the concept of sustainability, will occupy much of the remainder of this book.

Toward an Economics of Enough

The important final question of this chapter is as follows: What should the alternative be? Let us review the major conclusions of the chapter. The basic necessities of life have evolved into requirements *plus* wants, which have further evolved into wants as personal demands. Sufficiency and survival have evolved into wealth and acquisitiveness. In turn, the notion of unlimited personal acquisitiveness as a desirable trait has resulted in the sanctification of a system that favors maximization of consumption and production at any cost. Low production is bad, and high production is good, no matter what is produced. Consumption is assumed to be automatic as income is spent.

The quality of a good or service, or quality of life for the individual, is afforded little concern. Natural resources are to be used by humans to support this credo, regardless of what the effects of their depletion or degradation might be. Consequently, the natural environment as a whole is characterized as a virtual enemy of economic production.

Further, in the political sphere, a system of unregulated market capitalism is *vaguely* extended in the lexicon of civic discourse as the ideal. An unregulated market is the only system that even begins to fit with an ethic that sanctions unlimited growth through unlimited consumption. Without further exploration into the political implications, we say "vaguely" because it is much easier to invoke generalized public sentiment against government by simply chastising it as inefficient or calling for "less bureaucratic regulation and red tape" than it is to specify exactly what this means. On a point-by-point examination, polling and surveys show that citizens/voters tend to favor the protections offered by many of the regulations, including protection of the environment—making it dangerous to delve too deeply into the subject while on the campaign trail.

Alternatively, human motivation for material satisfaction would be better focused on meeting basic necessities and limited wants—a notion of sufficiency or "enoughness." Think about it. A value system that favors continuous acquisition leaves the individual in a perpetual condition of existential insufficiency and unhappiness. They never have enough. If *more is better*, then *enough* never comes. Such a person will end up less happy despite using more resources than someone who consumes fewer resources and yet experiences contentment. Incidentally, such a result—more satisfaction with fewer natural resources—ought to qualify as a classic definition of efficiency, which de facto becomes effectiveness as well. Ironically, it is the pursuit of the current unenlightened concept of *efficiency* that has resulted in such an *ineffective* system.

If one accepts the notion of sufficiency, and the idea that there might reasonably be an upper limit on the consumption of most goods and services, competition might yield more freely to cooperation. Hoarding thus becomes seen as unnecessary and so yields sharing. Quantity of goods gives way to quality of life. Civic attitudes would tend to evolve from simply wanting government protection from economic exploitation to a role for government as protector of the long-run health of the commons, as everyone's birthright, and a provider of basic support services for its citizens. More support will exist for acting collectively in such areas as education, public safety, parks, public health, environmental protection, and infrastructure provision, where involvement of the public sector is not only possible but also demonstrably *more effective* than providing these services privately.

In the following chapter, we turn to the topic of *production*, or how the producing sector of an economy responds to the necessities and desires of individuals to consume in order to thrive. Nonetheless, the important issues we raised under the topic of consumption will remain critical to this task. The economy is a system, and it can be neither studied nor understood effectively in any other way.

5

Production

The beginning section of this chapter might sound familiar. If the previous coverage of consumption begins with the *need for* economic output on the part of human beings under the mandate of securing necessities, this coverage starts with the other side of the coin—the *supply of* those necessities. How do producers produce? What do they assume? What are the systemic effects of how they behave? In what ways have the how and why questions changed over the years? These are the types of questions addressed in this chapter.

In light of the previous discussion of the theory of consumption, it is tempting to conclude that individuals, as consumers, have left the producers of goods and services with an onerous task. Given the standard assumption of unlimited wants, no matter how much consumers are provided in the way of goods and services, it will apparently never be enough. The theory, we concluded, implies that the consumer can never be satisfied—a notion assiduously stimulated by the advertising industry in concert with producers.

Original Intention: Meet Human Necessities

In a subsistence society, the job of the producer is straightforward: provide the material means to stay alive. In the ultimate version of a subsistence society, a one-person world, the consumer is also the producer. This is a situation economists are fond of calling "Autarky" or the "Robinson Crusoe Society." If people have to fend for themselves, they are responsible for all they consume or use. This, almost by definition, would be a subsistence economy, because accumulating wealth would have no meaning other than to secure leisure time apart from the effort of gathering the means of survival from the surrounding environment.

The economic dynamics of Robinson Crusoe do not change substantially if we expand the analysis to encompass a tribe or village. The emphasis is still on survival, albeit to that of the community and its members as opposed to just the individual. Here, however, we are able to incorporate the issues of specialization and division of labor. These terms represent the economic characteristics that form the cornerstones of Adam Smith's seminal treatise on economics, *Wealth of Nations*. Divide up tasks and prosper. Specialize and trade. It is useful to create a simple example.

Suppose there are two needs for basic survival in a small tribe or village: food and shelter. One person or family can provide all its own food throughout the year by working an average of 4 hours per day. Furthermore, it can provide all its own shelter by working another 4 hours per day. Thus, meeting all its own needs requires an average of an 8-hour day throughout the year. The same is true for any other family or individual. Thus, both work 8-hour days in order to meet basic needs and survive.

Next, assume that for some reason, one person finds he or she is particularly good at obtaining food—whether it is hunting, gathering, farming, or whatever. Perhaps a man or woman finds that he or she enjoys economies of scale by growing a bigger garden wherein he or she is both efficient and effective in the practice of weeding, watering, and harvesting. Assume further that a second individual enjoys a similar experience with providing shelter. That individual may specialize in construction, finding materials, using the appropriate tools, and so on.

As a result of these efficiencies, the first individual can produce enough food for both of them in 6 hours per day, and the second can similarly create enough housing for both in 6 hours per day. If the two of them decide to cooperate, the first can produce only food, the second only shelter, and then each gives half of their specialized production to the other. As a result, each will have his or her requirement of food and shelter met and will only have to work 6 hours per day—instead of 8 hours. This act of specialization and division of efforts will result in 2 hours more per day of leisure, possibly to nap or even draw pictures on the walls of caves.

Consider that hunter-gatherer peoples were the original affluent societies because they lived with few material possessions for hundreds of thousands of years and enjoyed lives that were in many ways richer, freer, and more fulfilling than ours are today. These peoples so structured their lives that they wanted little, needed little, and found what they needed at their disposal in their immediate surroundings. They were comfortable precisely because they achieved a balance between what they required and wanted by being satisfied with little.

With relatively simple technology, such as wood, bone, stone, fibers, and fire, they were able to meet their material needs with a modest expenditure of energy and have the time to enjoy that which they had materially, socially, and spiritually. Although their material wants may have been few and finite, and their technical skills relatively simple and unchanging, their technology was, on the whole, adequate to fulfill their needs, a circumstance that allows us to conclude some hunting-gathering peoples were the original affluent societies—not part of an ordained tragedy in which they were prisoners at hard labor caught seemingly forever between the perpetual disparity of unlimited wants and insufficient means.

The !Kung Bushmen of southern Africa, for example, spent only 12 to 19 hours a week getting food because their work was social and cooperative, which means they obtained their particular food items with the least possible expenditure of energy. Thus, they had abundant time (leisure) for eating, drinking, playing, and general socializing. In addition to which, young people were not expected to work until well into their 20s, and no one was expected to work after age 40 or so.[1]

With respect to leisure, there are two conjoined sides to the concept. We tend to think of leisure, according to Brother David Steindl-Rast (a Benedictine monk), as the privilege of the well-to-do. "But leisure," says Brother Steindl-Rast, "is a virtue, not a luxury. Leisure is the virtue of those who take their time in order to give to each task as much time as it deserves. . . . Giving and taking, play and work, meaning and purpose are perfectly balanced in leisure. We learn to live fully in the measure in which we learn to live leisurely."[2] This sentiment is echoed by Henry David Thoreau: "The really efficient laborer will be found not to crowd his day with work, but will saunter to his task surrounded by a wide halo of ease and leisure."[3]

Living leisurely was a trait of the Shoshonean People, who arrived in what is now Death Valley, California, about 1000 AD. The Shoshonean People were the seed gatherers of the desert. Much of the year they lived among the sand dunes in simple shelters of brush, where they harvested mesquite beans. But when the seed of the piñon pine ripened, they camped in the nearby Panamint Mountains for the harvest.

They also gathered what other seeds they could and used smooth flat rocks to grind seeds into flour. In addition to gathering plants, they hunted such small animals as rodents and lizards and ate adult insects and the grubs of beetles.

Although their tools were simple, the people possessed great skill. The ability of these people to find and utilize whatever foods the desert offered was the key to their survival.

The simple society of the Shoshonean People afforded two things that have so far eluded us in modern life—ample leisure time and the peace to enjoy it. Their free time was not devoted to improving their material standard of living as is ours, perhaps because that rung on the cultural ladder was unattainable in an environment permitting no cultural evolution, but then perhaps it would not have been perceived as a necessity or even a desire of life had it been possible.

The Shoshonean People thus lived in fullness within the context of the natural cycles of their environment and their lives. Their trust was founded on the natural law of replenishment, just like breathing in and breathing out. But while the environment provided a subsistence that allowed ample time for leisure, it also precluded the luxury of war, an activity that requires its own technology and a perceived abundance of resources to waste. When warlike tribes entered the valley, the residents just slipped quietly away and hid until the intruders left.[4]

However, the advent of agriculture brought with it both a sedentary way of life and a permanent change in the flow of living. Whereas the daily life of a hunter-gatherer was a seamless whole, a farmer's life became divided into home and work. While a hunter-gatherer had intrinsic value as a human being with respect to the community, a farmer's sense of self-worth became extrinsic, both personally and with respect to the community as symbolized by, and permanently attached to, productivity—a measure based primarily on how hard a person worked and how much that person produced.

These are telling examples, for by adjusting them and examining the various ramifications, we gain insights into the evolution of production as a process. To illustrate, it may be possible to produce much more of a particular product than is required by two individuals or families, and thus several or even many can be served. "I'll make the bow, and arrows. Running Bear, you and your sons do the hunting. Morning Star can gather edible plants. Smiles-a-lot can help tan the hides, and her sister can sew moccasin." With time, agriculture, and permanent settlement, these early skills evolved into the farmer, village cobbler, blacksmith, and tailor. Moreover, as the number and distribution of towns and cities, which focused human interactions, grew denser, the easier it became to specialize production and trade for necessities. This division of labor, in the form of relief from having to spend every waking hour in a mere survival mode, resulted in an ability to develop increasingly complex cultures.

The foregoing scenario leads to two prominent conclusions. First, the opportunity to produce and garner more goods than just food and shelter is clear. Different people could produce clothing, housewares, furniture, or even specialize within the arenas of food and shelter: "I'll produce chickens, you raise pigs, John can grow carrots, Sam can get wood, and Sue can do the cooking." Thus, the movement away from bare necessities toward a more sophisticated economy becomes inevitable.

Second, the notion of markets begins to materialize. Up to this point, the dominant assumptions about the nature of exchange have been primarily a barter economy. We each produce what the other also needs, and then we trade. Such exchange becomes cumbersome when a diverse variety of products is produced. The notion of the "middle-age" marketplace will work to a degree, and certainly has appeal as a community-building activity. But, even then, market prices, in the form of trading ratios, are virtually required. The notion of money as a medium of exchange, and in support of a less personalized market process, cannot be far behind.

Modern economists look down their noses at a system of barter compared to a price-based market exchange. Barter is simply unable to support the level of economic activity necessary to keep a modern, industrialized economy healthy. This said, traditional economists, with their single-minded focus on efficiency, consider losing the interpersonal social capital of community marketplaces as more than offset by the sheer increase in the volume of output possible with the completely impersonal, modern market, where all that matters is the price and the volume of transactions.

As a side note, and perhaps optimistically, we assume that people yearn for more than impersonal marketplaces. Many people are increasingly finding that their roles as strict price-seekers in the mass market (such as the modern, big-box discount store) are increasingly unsatisfying and distasteful, even given the welcome low prices, as they struggle to meet basic needs in an economy that is failing them. A desire to repersonalize the link between consumer and producer underlies the rising popularity of local food movements and Saturday markets. Included is the desire to minimize the requirements of transportation and distribution and the wasteful energy expenditures that accompany them, and to restore an element of community into the act of meeting basic, material necessities.

The Goal Has Been Unlimited Production

These observations raise an important point. Notice that the emphasis for production theory in the evolution of economic thought has shifted subtly away from a focus on the individual meeting his or her life's requirements, and has shifted toward the producer or business. Diminished is the attention given to the need to support the *person,* and increasingly the focus is on keeping the *business* viable. Obtaining the goods and services needed for survival, or even a comfortable life, for the individual or family receives decreased attention. Maintaining an acceptable, and preferably growing, volume of sales for the producing entity becomes the center of attention.

Thus, economic health is seen as keeping business viable, as opposed to primacy on ensuring that individuals and families are well supported. The economic problem has shifted from *supporting individuals* to *supporting businesses.* As an example, in the current near-disastrous economic climate, marked by palpable human sacrifice and suffering, the political pressures in Congress seem to focus on incentives for business (lower interest rates, capital gains tax cuts, and so on), rather than on bolstering human welfare through such direct measures as extending unemployment benefits. Raising production in support of businesses apparently enhances political capital more than does bolstering the affordability of available goods and services in the support of families.

To be sure, the realization is still tacitly present, and occasionally acknowledged, that the seminal reason for businesses to exist is to produce for consumers. Humans have wants and needs, and this is the reason an economy exists. Productive enterprises, using whatever technological means are available and appropriate, need be seen simply as a mechanism or delivery vehicle for accomplishing material well-being.

Somewhere along the way, however, the emphasis shifted. Profit, for instance, was initially the reward for satisfying human necessities and some wants. If a producer (in earlier times, perhaps, merely another human being in the role as an artisan or craftsman) provided something seen by others to be of value, the producer could be rewarded with monetary compensation, which would allow them to similarly meet their own desires. In fact, the concept of *profit* or *loss*, in the sense that the producer either does not cover costs or covers them and has a significant amount left over, is a post–Industrial Revolution idea. What's more, this idea developed only with the growth of larger, impersonal producing entities paying salaries to the workers, as opposed to a single proprietor who simply lives off the returns for services, whatever that amounts to. For all practical purposes, profit and the corporation evolved simultaneously.

As a final point, the behavioral assumption of unlimited wants on the part of Rational Economic Man is nonetheless retained. Therefore, despite a transition within economic theory over the years to focus on the producer rather than on the consumer, the task the producing entity inherits borders on the impossible: *Produce as much as you can of anything you can sell, and it will still probably not be enough.* That is, provided advertising has done its job on instilling a constant sense of lack and its corresponding dissatisfaction—a mandate proposed by 20th century economist and retail analyst Victor LeBeau, who wrote in the *Journal of Retailing* (Spring 1955):

> Our enormously productive economy demands that we make consumption our way of life, that we convert the buying and use of goods into rituals, that we seek our spiritual satisfactions, our ego satisfactions, in consumption. The measure of social status, of social acceptance, of prestige, is now to be found in our consumptive patterns. The very meaning and significance of our lives [is] today expressed in consumptive terms. The greater the pressures upon the individual to conform to safe and accepted social standards, the more does he tend to express his aspirations and his individuality in terms of what he wears, drives, eats—his home, his car, his pattern of food serving, his hobbies.[5]

As we will see both later in this chapter and throughout the book, this cultivates a sense of lack, of never having enough to feel secure, and has fundamental implications for both our environment and our values.

Reconciling the Differences

How do we assess the evolution of the completely interpersonal economics of the early tribe or village to the modern, completely depersonalized economics of globalization?

Let us be clear as to what has been lost in this evolutionary process. It is human dependency. People depended on one another economically to their mutual benefit. I will produce your shoes and you grow food for me. You weave clothing and I will construct your home, and so on. It is the notion that each of us is at once a producer and a consumer. By this, we mean that everyone has a gift to offer society as a whole in that all gifts are complementary, whereby no one's gift is more or less important than anyone else's, all are required for the mutual benefit they offer. In this way, we have depended on one other for our survival since time immemorial, and for any level of prosperity above subsistence that might be attainable.

Put more succinctly, what is lost today is a *sense of community* that overlays and integrates with the economy. Clearly, anything approaching the full restoration of this feature is impossible in a world rapidly approaching seven billion people. It takes much more awareness of the human condition than most of us can muster in order to feel a sense of community with a Chinese worker in a factory producing some trinket for export—often at the cost of an outsourced American job. But, as we contend in the concluding section of this book, there may be positive steps we can take.

A question arises. Suppose an American consumer, or perhaps a consumer group, observes with alarm the degree to which the dramatic imbalance of trade with China is embraced—even perpetuated. We import over 10 times the value of goods from China than we export to them. People often comment that it seems like we do not produce anything in the United States any more, even when our economy is visibly in recession. Suppose further that we were now to embark on an aggressive campaign of "Produce American" and "Consume American."

Abstracting from the financial, macroeconomic, and foreign trade policies that might be required to achieve this, the question becomes as follows: Would this represent a healthy, return step in realizing the economic futures on which Americans are interdependent, or would it be a form of self-centered "beggar-thy-neighbor" nationalism leading to increased international tensions and perhaps even more world economic inequality? ("Beggar-thy-neighbor" is an economic expression describing a policy of seeking benefits for one country at the expense of others.)

The question is largely unanswerable—even though an overall response is *probably aspects of both*. However, the important point is not the answer to this hypothetical situation, but the fact that we are driven to raise the question in the first place. It underscores the fact that globalization actually represents the ultimate depersonalization of the economic process.

For our traditional economics, and especially David Ricardo's venerable theory of *comparative advantage*, it should represent the crowning achievement. Finally, goods are all produced in the cheapest place possible in the world, and trade (read *mutually beneficial exchange* in the textbooks) is ubiquitous.[6] There is, however, one dramatic difference between the modern world and the one imagined by Ricardo. Namely, the dominant institution is no longer the *nation* but rather the *transnational corporation*. This demonstrates once again, but on the ultimate global scale, that production has finally dominated consumption. To the powers that be, both private and public, the economic imperative for success is seen as keeping corporations viable and profitable.

In truth, we created a world of dramatic inequality wherein producers cannot afford to consume what they produce, and consumers cannot be hired to produce what they consume. This world is the polar opposite of a smaller-scale, locally sustainable economy. It features the ultimate alienation between the producer and the consumer. Such a world cannot survive. And facilitating this perilous economic journey is the multinational corporation, which pays subsistence wages in the producing countries and charges all the market will bear to consumers in the industrialized countries. In the process, they reap the difference as profits for the corporate elite, dramatically increase inequality, and violate the environment as a final legacy. Our conclusion is harsh, but unavoidable: *Modern economic theory is a tragedy.* We leave it there.

We cannot conclude our broad-brush coverage of production theory without exploring in some depth an accompanying concept that, over the years, has been commandeered to play an instrumental role in its evolution. This is the concept of *productivity.*

The Concept of Productivity

Productivity is defined simply as the output per unit of time for a particular input. As a term in economics, its most common application is to human labor. An individual, or a worker, is productive if he or she creates something of value through his or her efforts within a certain amount of time. Thus, the term has come to be represented, in its most common form, as *output per worker per hour.*

The goal of a successful economy is commonly seen as achieving *increased productivity* over time. Who could possibly find fault, in theory or practice, with something leading to "higher" productivity? After all, the word implies the ability to realize a payoff or reward for effort expended. This is the essence of a successful, free-enterprise economic system—value received for value expended.

If, for example, an individual possesses talents and skills that are of value in producing something, then productivity represents the ability to employ these talents and skills in the labor market (i.e., work for a living). An increase in skills, say through additional education or training, leads to a gain in productivity, which ideally results in the willingness of an employer to increase the compensation for these skills by awarding a raise in pay.

Therefore, increased worker productivity ideally leads to a higher salary, which leads to a greater ability to command consumer goods and services, and—remaining within a totally materialistic frame of reference—a higher standard of living. Consequently, productivity supports economic satisfaction in a series of often-complex relationships, and increased productivity supports not only higher wages for the individual over time but also economic expansion and growth for the broader economy.

So far, this is standard modern industrial thinking. Productivity in the labor markets leads to purchasing power in the final product markets, and the circular-flow economy remains in balance, and even expands over time. But the plot thickens when we remind ourselves that human labor is not the only valuable resource an employer seeks to acquire. There are other factors related to production, such as land, capital, and various raw physical materials. In its purest form, the discipline of economics is designed to focus on the productivity of all inputs (output per acre, productivity of capital, efficient use of land, energy efficiency, and so on), not just on labor. We pursue this thread at greater length in the final chapter of this section, which deals with the macroeconomic system.

It is our strong belief that many of the perceived threats to the environment and the carrying capacity of the globe stem from a failure to acknowledge this reliance on all resources and consequently concentrate single-mindedly on *productivity per worker*. In order to explore the implications of this narrow focus, let us shift our attention to the economic notion of capital, which for our purposes can be broadly defined as factory, equipment, and machinery.

Economic theory holds that the components of production, such as capital (the machine) and labor (the person running the machine), combine in some way to create output. The exact manner in which they combine is defined by the state of technology. And the theory contends that each factor should be compensated according to its actual contribution to the final, measurable output of the product.

Any adjustment, technological or otherwise, which increases the productivity of one component, decreases the productivity of the other. The productivity of labor and the productivity of capital each operate in the opposite direction of the other, because in economic theory, they are analyzed as a mix (i.e., the capital/labor ratio).

Consequently, a resource that enjoys increased productivity, for whatever reason, should receive higher compensation in the form of an increased share of the incoming revenues. This, of course, implies a decrease in the relative share of the incoming revenue going to the other component or components. In the absence of growth and expansion, what one gets, the other does not.

The law of diminishing returns holds that productivity of the last productive component to be added (either capital or labor) will normally drop the more there is of it. In other words, if a business hires more workers, the incremental increase in productivity of the last hired worker drops when compared with the productivity of the other workers, *assuming a fixed amount of capital*. This principle holds true in either direction. As more of either, capital or labor, is employed, the incremental productivity of that more abundant component drops as the incremental productivity of the less abundant component increases. Therefore, in order to increase the productivity of one factor of production, it must be accorded a proportional increase in the other factors of production.

Imagine a simple situation that will exemplify all this. A farmer owns some land that he wants cultivated. He needs workers (labor) and shovels (capital) to accomplish this. Initially, he hires one worker and buys one shovel, and a certain amount of work gets done per hour or per day. If he considers hiring another worker but only has the one shovel, daily output will presumably increase somewhat, because one worker can work furiously and then rest for half the time while the other intensively operates the shovel. The shovel is never idle. Again, output may rise due to the second worker, but it will certainly not double, because they only have one shovel between them. The value to the farmer, in terms of willingness to pay the second worker, is thus less than that of the first worker.

On the other hand, suppose the farmer considers buying a second shovel but hiring no more workers. The initial impression would be that the marginal product of the second *shovel* is zero, because shovels cannot operate themselves. However, there may be some argument for a small positive benefit for the second shovel—the first one might break, and the second as a spare would allow production to continue without a work stoppage.

Translating this back into the practical world of the farmer, he or she clearly needs an appropriate mix of all resources. Financially, the farmer will readily pay for the first worker and the first shovel, because any production depends on having some of each. If the farmer considers increasing the amount of one of the inputs without increasing the other (i.e., becoming more labor intensive, or more capital intensive), the farmer will not be willing to pay as much for the next unit of that resource, because its contribution to total production will be visibly less than that for the first input employed. In fact, what history tells us a "growth-oriented" farmer might do is hire five more workers, buy five more shovels, and seek to expand the whole operation. Then the constraining input might become the availability of land.

As a conclusion to this chapter, we combine all these points of standard economic analysis to construct a Capitalist Scenario, which serves to point out a core systemic irrationality that follows from the incorrect application of the economics of production in the corporate real world. This scenario is best depicted in straightforward bullet points:

Capitalist Scenario

- A resource must be paid more if it becomes more productive.
- A resource becomes more productive if it is given relatively more of the other resources to work with.
- In their quest for higher output, owners of capital seek consciously to automate—that is, to increase the proportion of mechanical labor to human labor.

- This means that capital (the machine) is becoming the *less* valuable incremental input, and human labor is the *more* valuable input.
- Therefore, labor should be paid more and capital less.
- Although profits rise temporarily after automation, businesses perceive the opposite—that capital is the more valuable resource.
- They, therefore, continue to automate and further replace human labor with capital, and seek to pay workers even less.
- This leads to a saturation of the economy with the less productive input (capital) and to unemployment and underpayment of the more productive input (human labor).

This scenario leads to—and depicts—the essence of the modern tragedy global-corporate-capitalism has become. It is not a basic failing of economic theory, but rather of business practices employed in violation of that theory. We temporarily leave the scenario at this point, but note that the unfortunate ironies, which it puts on display, set the stage for the final two chapters of this section: Chapter 7 on distribution and Chapter 8 on the macroeconomy. We will then revisit the Capitalist Scenario in deriving our final conclusions about the current state of economic theory and practice. First, however, we must take a side trip through the knotty thicket of externalities and economic theory.

Endnotes

1. John Gowdy (ed.). *Limited Wants, Unlimited Means.* Island Press, Washington, DC. 1988.
2. Brother David Steindl-Rast. *Gratefulness and the Heart of Prayer: An Approach to Life in Fullness.* Paulist Press, Ransey, NJ. 1984.
3. Henry David Thoreau. http://quotationsbook.com/quote/42321/ (accessed on November 23, 2010).
4. Our discussion of the Shoshonean People is based on: William D. Clark. *Death Valley, the Story behind the Scenery.* KC Publications, Las Vegas, NV. 1981.
5. Victor LeBeau. Price competition in 1955. *Journal of Retailing* Spring (1955):5–10, 24–44.
6. Comparative advantage. http://en.wikipedia.org/wiki/Comparative_advantage (accessed on November 25, 2010).

6

Externalities

In this chapter, we address the area of economic theory known as *externalities*, which can be thought of as a subset of economic theory. Externalities, in these cases in a competitive market, are those unaccounted-for costs or benefits that accrue to society as a result of producing or using a product or service. To illustrate, manufacturing that causes air pollution exacts costs on the whole of society for generations but is not accounted for in the economic formulation of how the product is priced, so the product may be overproduced in terms of its benefit. On the other hand, too little economic attention may be afforded education because the wide-ranging, long-term, societal benefits of the educational process are not taken into account by those involved in the educational process as a whole.[1]

Politics, Economics, and Externalities

In point of fact, externalities are often treated as an irritant, an unwelcome relative, because they necessarily lead to awkward, embarrassing qualifications and exceptions to the major elegant results of the core features and conclusions of economics as a discipline. On the surface, therefore, the notion of *externalities* may not seem to hold the same importance as more central areas of economics, such as consumption, production, distribution, and macroeconomics. We disagree wholeheartedly.

There are important, practical, and controversial real-world issues, many of which affect us every day, such as global warming, where the role of externalities *is central* to the controversy. A sampling of some of these issues begins to look like the evening news—or perhaps a *good* political campaign.

Internationally, global economic crises are exacerbated by the practice of ignoring externalities. The immediate result is to underprice valuable natural resources, which leads to waste, overexploitation, and even the subjugation of indigenous and poverty stricken populations. Certainly, this lack of systemic attention increases worldwide tensions by widening the rich–poor gap within and among peoples of the world and, thus, nations. The rich and powerful have always had the ability to control the market through the corporate form of organization, thereby largely avoiding unpleasant personal effects of economic activity by shifting the negative burden of cost to those who do not benefit directly from the market transactions in question.

Many costs, properly termed *externalities,* are never acknowledged in the determination of the market prices and quantities of all we consume. This includes the social burden for workers who become injured or ill, or for whom little health care is provided. Substandard working conditions or inadequate pay for valuable services provided are not included in the price.

Firms extract resources from public lands and pay fees that are a ludicrous fraction of their true value; yet they often receive subsidies for doing so. Moreover, many public resources—other than those in question—are depleted, polluted, or destroyed in the process. Normally, these are not counted when the costs are added up and rolled into a price. In fact, any proposed regulations that would either limit the damage or force compensation are vigorously opposed and denounced as inefficient.

Within business practices, taxpayers support the communications and transportation systems. Huge public relations, lobbying, and advertising campaigns are supported at public expense—often to convince us that the damage is negligible. Overly generous salaries and fringe benefits for the executive decision makers impact innocent third parties through their tax deductibility. What's more, we are often *paying to be polluted* by industry.

In summary, the factors that fall under the heading of externalities compose many, if not most, of the perceived failings of a market-based economy. It is patently obvious that factors, which the discipline terms *external, are totally to blame* for the fact that we face potentially disastrous and irreversible impacts to the global climate.

Most economists would argue that by acknowledging externality theory, the body of method has the capability of addressing these all-important issues and, in effect, of proposing effective solutions. But it does not seem to happen in the real world. In the *real* world, both our economic and environmental fate as a culture may well depend on how we choose to deal with externalities. We can choose to use economics as a tool to either help or harm our efforts to foster social-environmental sustainability in the years, decades, and centuries to come.

As we will see in this chapter, jargon is a serious problem. But the inadequacy of the language is also a glaring indicator of a deeper and more serious problem: The practical effectiveness of any body of analysis must be seriously questioned if it refers to the world in which we live and upon which we depend for our existence as "external." This is the ultimate irrationality.

Understanding the Language

The topic of externalities in economic theory provides a graphic insight into the failure to use the discipline of economics to deal effectively with many of our currently perceived societal problems. Let us be clear. With most of our observations concerning the exploitation of supposedly solid economic doctrine, the problem stems from overzealous or inappropriate use of the theory, and not directly from the theory. Practitioners, businesspeople, and those who purport to be boosters of a local or regional economy often misuse economic concepts if they are perceived to be supportive of their own particular policy interests. This behavior can give the discipline an undeserved bad image, when in actuality it is the villainous application of the principles that is problematic.

This is not necessarily the case with the theory of externalities, because the structure of the accepted dogma invites nothing less than the tragic application of that theory. The place to start is with the jargon. Depending on the theoretical application of economic analysis, several terms to define externalities can be used, which, under certain conditions, are synonymous with *externalities*. These include:

- External benefits
- External costs
- Spillover effects
- Market leakages
- Market failure
- Imperfect property rights

Notice that anything directing resources in directions other than through markets is labeled a *spillover, failure, imperfection*, or *leakage*. Inherent in these terms is a profound implication. Namely, there is the clear presumption that all worthwhile, or "valid" economic activity is assumed to occur within markets. In effect, the only socially optimal resource allocation (such as the distribution of wealth, goods, and services) is that which is bestowed on people and businesses through the activity of private markets. The implied premise can be stated even more clearly: *Free-market capitalism is unambiguously the best way to organize an economy.*

This statement is critically important and clearly values based. At first glance, it may seem to involve an unwarranted leap in logic. To those accustomed to viewing themselves as favorably disposed to capitalism, making this assertion will seem to involve a sweeping and unwarranted accusation. That said, we are not taking sides on the values-based ideology of either pro- or anticapitalist arguments. Rather, our purpose is to dissect the implications of the discipline *as it is structured*. Accordingly, the assertion calls for considerable explanation, additional terminology, and cogent examples if it is to be substantiated. That is the purpose of this section.

The Nature of Markets

To begin, an exchange economy has certain characteristics. The term *exchange economy* refers to an economic order, wherein the bulk of economic activity occurs through the free (unrestricted) buying and selling of valuable goods and services. The important point for most economists is that all transactions on the part of sellers (suppliers) and buyers (demanders) are voluntary. All purchases and sales of goods and services, including selling one's own labor by accepting a job, are entered into voluntarily and presumably with the expectation of improving one's economic status or well-being. It is assumed that an individual will be happier, or better off, after the transaction, and that a business firm will be more profitable. Profitability on the part of a firm is, of course, the equivalent of consumer satisfaction for the individual. It is that which the economic actor seeks to optimize. In this manner, pursuing one's own self-interest through market activity (whether a buyer or seller, a demander or supplier, a business or an individual) can be said to optimize well-being for the society as a whole. In other words, pursuing one's own self-interest is what is best for the entire system—the convenient and happy upshot of Adam Smith's famous "Invisible Hand."

Before going further, we can note parenthetically an initial *value bias*. The driving forces behind this notion are freedom to act without constraints of any kind (from either government or private interests), along with the presumption that the goal is the maximization of one's own self-interest (lack of altruism). It becomes clear that both a conservative and libertarian philosophy would be attracted by such economic premises.

This helps explain why the concept of *freedom* is so often touted as support for *free markets*. In truth, freedom is much more than just a free market, but apologists tend to equate the two, because that allows for seemingly strong cultural or social support for their ideological biases. Of course, absolute freedom never exists, because everything is constrained by its relationship to everything else (including every market), which means that we are always operating under constraints of one type or another. Much social strife, especially in these days of *culture wars*, could be lessened if everyone maintained their conscious awareness of this irrefutable fact.

We need dwell no further on the often-heard accusation that economics ends up an innately conservative discipline, but an important point, made earlier in this work, should be reiterated. The contention that economic analysis is value free is a complete falsehood. To be sure, the rules of mathematics and logic are predominant once one chooses to enter into the system and employ the analytical tools (e.g., to become an economist, or perhaps even a business person). Everything follows from facts and assumptions, and the methodology becomes all technique. However, the all-important value judgments are made in advance by choosing to accept unrestricted wealth and greed as the operating framework. These choices *are ultimate value judgments*, and need to be recognized as such.

Imperfect Property Rights

Returning to the main argument concerning externalities, there is an important implication for the notion of property rights. In essence, economic activity is simply the buying and selling of property rights. A buyer of a car or a head of lettuce is giving up the property rights to money in order to secure the use rights to drive the car or eat the lettuce. The auto dealer or the grocer does the reverse in accepting the right to use the money (the sales proceeds) by giving up the rights to the car or the lettuce. Exchange has occurred. This is an example of final-product-market activity. The same holds true in resource markets wherein the buyer and seller relationship is reversed. Thus, an individual sells to an employer the right to his or her time when he or she accepts a job.

Markets work well to the extent that these property rights are exact and strictly adhered to. In other words, the right to use the good or service must be clearly and unambiguously owned by the seller and thus able to be consumed or used solely by the buyer.

If these conditions (exact ownership, free transferability, and exclusive rights of use) all hold, convention says the system has perfect property rights. The meaning of this for externality theory is fundamentally important: *Only the buyer and seller of a good or service are affected by the production, sale, or consumption of a good or service. In other words, externalities do not exist.*

In this ideal world, it could be said that a free-enterprise market economy is the best possible way to organize the meeting of material necessities of both individuals and entire societies. However, this is a very restrictive condition—a fact ignored in the much-simpler world of Adam Smith. In truth, no property right is innately perfect. Our actions—especially in this increasingly crowded world—affect others. In the remainder of this chapter, we do our best to demonstrate that pursuing an economic state of affairs, which clings to this "zero-externality," idealized methodology, has resulted in serious crisis conditions for both the economy and the environment.

Proceeding through Example—The Paper Mill

An effective way to proceed in elaboration of the above assertions is through an extended example, which we use to elucidate both the terminology and the underlying basic principles. Any reasonable example must include seminal characteristics. First, a typical act of production or consumption (i.e., resource allocation) is identified. Second, the pattern of resource allocation results in scientifically identifiable technical or physical impacts, in addition to the obvious economic flows as the product is produced, sold, and consumed. Third, potential misallocation of resources is identified, as well as the policy implications, including potential legal liabilities, regulation, and political actions.

Assume the existence of a paper mill in a small to medium-sized town located in a timbered area of our country. The factory, located in a town on a river, produces and sells paper, employs people, and creates revenue for the owner of the mill. In the course of production, as has obviously been the case throughout the actual history of this industry, smoke and chemically laced fumes escape into the air, and toxic effluents enter the adjacent waterway.

Our purpose is to facilitate both a technical and value-based understanding of the problem's implications, which arise from the use of externality theory, as it is structured. In other words, this effort is intended to critique *economics*—not to analyze or solve the environmental issues being raised.

First, we must identify the parties to the bare-bones *economic* transaction. There are four broad categories, including (1) the owner or owners of the mill, (2) the suppliers of all the inputs necessary to the production of paper, (3) the consumers or final users of the paper, and (4) the sales and distribution network. Each of these requires a bit of elaboration.

There may be one owner, or more likely a corporation, whereby any stockholder can legitimately consider himself or herself to be an owner in position to profit from the production and sale of paper. The suppliers include those who log and haul the timber to provide wood pulp, energy, chemicals, or any other raw material necessary to the paper-production process. Certainly, the major supplier of inputs is the labor force, or the people who work for the paper company and therefore earn their livelihood from it. Consumers seek to purchase and use the paper and, of course, expect to pay for it at some store. Finally, there is a broad network of those who package, transport, distribute, and market the paper, down to the local store that stocks it on the shelves (along with many other products) and deals directly with the customer.

In summary, all these economic actors have a stake in the existence of paper as a product. Some make virtually all their income, some make a portion of their income from the production of paper, whereas others simply gain satisfaction through its purchase and use. Each is exchanging property rights as they become involved in the various markets: valuable inputs for money, labor for money, services for money, or money for the product. In an ideal world, as we saw in the previous section, all these property rights are perfect. All of the transactions, including the choice to work for the paper company, are voluntary and thus self-maximizing, and no one is affected outside the group of these identified parties.

In the real world of our assumed paper mill, however, this is clearly not the case. Other individuals or groups are affected. Residents of the town and its immediate environs experience the bad odor from the smokestacks, and perhaps even become ill from chemical sensitivity. Anyone who uses or recreationally enjoys the waterways, whether near the town or further downstream, can experience a lack of satisfaction or enjoyment as a result of the water pollution. This could include the inability to fish, swim, and boat; to use the water for agriculture and industry; or perhaps even to use it directly for municipal use. Many externalities are present.

Here, a personal experience is apropos. I (CM) was once on a citizens' advisory council for a large pulp and paper company that discharged its chemical effluence into the river from which my hometown draws its drinking water downriver from the mill. When I asked the mill foreman what would happen if I took the mill's intake pipe and placed it below the effluent pipe, he replied as follows:

Mill Foreman: We'd have to close the mill.
CM: Why is that?
Mill Foreman: Because the water's too polluted to run the machinery.
CM: And you expect me and the other people of my hometown to drink it.

At first, my only reply from the foreman was a blank stare.

Mill Foreman: The answer to pollution is dilution.

> This was a patently ridiculous response, however, because the ocean has no outlets through which to flush itself, so dilution is impossible. Moreover, all the fresh water on Earth originates from evaporation of the ocean, thereby further concentrating pollution in a continual, self-reinforcing feedback loop, which affects everyone worldwide who relies on the ocean for such things as food.

In short, many parties experience a cost, even though they are not one of the "economic actors" identified above. Here, the terminology becomes important. They bear a cost, even though they receive no compensating benefit. On the other side of the coin, the market participants enjoy a benefit for which they bear no cost. In the antiseptic language of the economist, *resources are misallocated.* Making sure that paper production does not foul the air or that noxious chemicals do not pollute the waterways is a task that should properly fall to the industry—including bearing the costs of accomplishing that, whatever they might be.

For example, suppose there exists a (moderately expensive) technology that would render the air completely clean and leave the water unchanged from the original condition. The options for bearing the responsibility of paying for these include suppliers of inputs receiving less (including lower wages for the paper mill workers), profits of the owners decreasing, and paper being sold for a higher price. Choices among these options depend on market power and market conditions, as well as the will and power of government, which are of no consequence here. Nevertheless, the economic fortunes of all the direct-market participants would be predicted to decline.

Now it is time to bring into the picture economics as methodology. It enters into this example first by acknowledging that a misallocation of resources has occurred because there are uncompensated costs and benefits enjoyed for free. If you get something of value, you should pay for it, and if a cost is imposed on you, you should somehow be compensated.

Further, it is said that the market has failed, because it failed to contain all the legitimate costs of production within the domain of just the private market participants. Economics, as a discipline, might also contend that an overproduction of paper has occurred as a result of the price of paper being too low to include all private costs of production. The economic response, which borders on an exercise in circular logic, is that the *externality should be internalized*. The external social costs borne by the breathers of air and the users of water should be somehow transferred to the direct-market participants—so far, so good.

Here the plot thickens, however. There is a wide range of real-world factors that complicate the attainment of the simple-sounding process of "internalizing the externality." First is that the "benefits" of polluting, in the form of higher wages and profits, as well as lower paper prices, in no way have to match the costs of dirty air or water. The profit the factory owner makes has no quantitative relationship to the health costs of a local resident with emphysema, or to a young family that can no longer enjoy swimming or fishing in the local river.

A second issue stems from the fact that in the early stages of the industry, the impacts appeared to be small, isolated, and self-correcting in terms of ability of the (otherwise clean) air or water to assimilate the pollutants. This untested appearance led to the assumption that air and water were "free goods" for the taking, so no one was really affected by the pollution. Thus, historical convention resulted in a willingness to ignore the impacts.

In effect, as we deduced during the more recent era of environmental concern, this uncritical assumption bestowed the ownership of air and water to the industry in the form of private property, as opposed to the citizens at large as part of the global commons, which is everyone's birthright. The process of wresting back these ownership rights through legal action has had a long and tortured history in the courts. There is a huge difference between the public saying, "The air and water belong to all of us a part of our birthright, and you must leave it as clean as you found it for the sake of all generations," and the industry saying, "I have always used the air and water in production, and I therefore own the rights to them, and you must compensate me if I am to stop current practices." When such issues get to the courts, or even to the point of some jurisdiction considering pollution-control regulations, it normally becomes a matter of ownership of property rights. This is the only way our courts can operate and is a fundamentally important area of interface between our economy, citizens' birthrights, and our legal system.

Another reality is that many of us wear more than one hat in these issues. Owners of the firm might also like fishing and can afford to do so in some clean, exotic place. Workers at the plant could suffer from airborne pollution. We all breathe air, drink water in one form or another, and use paper. I may enjoy water sports and yet have real concern for my neighbor, who works at the paper mill. In this confusing array of constituencies, economics has tended, in the interests of growth, material gain, and supporting an ever-growing population, to support pushing ahead with industrialization, despite the fact that many of the impacts remain unknown. The ethic has been as follows: "If you have no conclusive proof that we're doing damage, you must allow us to continue."

Drawing Some Conclusions

Unfortunately, there is a multitude of situations, perhaps not quite as graphic as a paper mill, which display these characteristics. The result is often continued social and cultural strife, frequently within communities, even after scientific evidence has uncovered substantial negative impacts. The forces of the status quo even sponsor countervailing scientific efforts, either to prove the contrary or merely to cloud the issue by deeming the existing data as flawed and thus "bad science," thereby hoping to continue business as usual amid the confusion.

The overall point is that economics, as a body of analytical method that supposedly supports public and private policy, almost universally comes down on the side of a progrowth, business-as-usual status quo. In doing so, it rarely concedes the validity of either biological or physical science—even when the data are clearly irrefutable, such as the measurable melting of glaciers around the world. Sometimes this will be through its analytical methodology and other times merely through jargon. Some hypothetical statements that exemplify this point include the following:

- The market has failed. We must fix—jump-start—the market, because, above all, we must not give up the notion that markets are the absolute best way to organize our society and allocate resources.
- Economic activity supports workers and delivers the products we all need. We must either ignore some incidental externalities or somehow compensate the few losers, because society would be unwilling to do without the goods and services.

- Government regulation is inefficient and inexorably misallocates resources compared to privatization or the activities of the private sector. Any internalization of the externalities should occur simply as a result of negotiation between the affected private parties. There may be no such thing as a "public interest."

These, and many similar statements, are examples of mindsets to which the discipline of economics would lend support—or rather with which practitioners and vested interests could use economics to avoid change. It is becoming increasingly clear, as a result of a preponderance of scientific evidence, that allowing modern industrial society to proceed unabated in its current directions may be little short of disastrous. What can be said of a discipline that inexorably courts disaster?

Some fundamentally important points can be made from the joint vantage point of economics and the discipline of ecology. An enlightened economics—which informs the public as well as the private business sector of its options for rectifying the problem—must take into account all the information that ecology can offer. First, and a fundamentally important point for this entire book, is that the full extent of environmental consequences is almost impossible to know. For example, our hypothetical paper company may be measurably polluting the water downstream, but what of the trace chemicals that accumulate downstream and into the ocean over time? What if the cumulative effects of all such paper mills (and similar manufacturing activities) are inexorably rendering the oceans unfit for animal life?

In fact, I (CM) asked both the company's chemist and the Department of Environmental Quality's chemist what would happen if the chemicals introduced into the river and the ocean were to recombine with chemicals already in the aquatic system. Both said they had no idea but pointed out that most of a given chemical compound is inert. In reality, however, nothing can be inert in an interactive system—such as the entirety of our universe.

Therefore, simple logic, as well as sound economic theory, would dictate that the processes should be stopped immediately. The costs of being unable to feed the planet would be virtually infinite and, by all odds, far higher than the benefits of using paper—no matter how much we enjoyed it. Our psychological marriage to economics in its current form would not only disallow that but also would favor doing the opposite.

Facing Uncertainty

The most important question facing any contemporary society wanting to embark on a path of greater social-environmental sustainability may well be as follows: How do we act in the face of uncertainty? The discipline of economics, whatever the details of its methodology, has unerringly come down on the side of continual growth—despite any and all negative consequences. Consciously or not, it has supported proceeding unabated until major, negative impacts are proven. If we are to survive as a viable society with an attractive quality of life, an important reform must be to change that.

Economics must embrace the uncertainty and act in the interests of long-term, social health instead of short-term, materialistic gain. Economic policy and scientific knowledge, especially in the face of major probabilistic uncertainties, such as those surrounding global climate change, must be aligned.

Full life-cycle cost accounting, even of the difficult-to-measure intangibles, must become a rigidly held norm. To date, professional lip service is afforded this goal, but little effort has been made in actually applying it. One effort is the environmental impact statement, and it is enlightening that development interests often ridicule this tool as a needless and cumbersome blockade to progress. The fields of economics and ecology must, in effect, act as one. They are, after all, inexorably linked in cause and effect—the consequences of which form the legacy of *all generations*. Moving toward this outcome is a major purpose of this book.

Endnote

1. Externality. http://en.wikipedia.org/wiki/Externality (accessed on November 13, 2010).

7

Distribution

Textbooks on economics regularly recite the axiom that an economic system answers three basic questions: What? How? and For Whom? In meeting the material requirements of humans, these abbreviated questions translate into the following: What do we produce (composition of goods and services)? How do we produce it (productive arrangements and technology)? And for whom is it produced (who gets to consume the output)? Any economic system, whatever its particular characteristics, is supposed to answer these questions if it is to qualify as an *economic* system.

In this second section of the book, we essentially covered the first two of these questions commensurate with the discussion of production and consumption, without saying it in so many words. Responding to the third question—who gets what—is the all-important task of distribution theory in economics. It is the most important question *and* the most problematic.

The Question of Who Gets What

The most striking differences among economic systems, and the most strident controversies, occur in addressing questions of distribution. After all, answers to the "what" questions lean to whatever people want and need, in both a survival and cultural sense. To be sure, people from industrialized nations with wealthy economies demand a wider range of consumer goods, whereas people living in nonindustrialized nations with poorer economies can afford fewer "luxury" goods, as opposed to more basic necessities. Even then, especially in a private-property market economy, those questions largely answer themselves. Furthermore, modern-day questions as to *what* to produce may have little relationship to the actual requirements of citizens in a particular economy, given the intense focus on production for export and trade. As such, the chosen products in any given economy—especially the less industrialized—are likely to be whatever the particular, specialized resource base allows to be exported and sold somewhere.

For example, Ecuador extracts and consumes oil—not because its people need a great deal of it, but because there are large amounts of petroleum reserves in Ecuador, and wealthier nations are willing to pay for it. Consequently, the "what" question of production is often dictated by the global imperative, especially for nonindustrialized economies, which are smaller, poorer, and less self-sufficient than industrialized economies. There is, however, a trade-off for the producing country—namely, great degradation, not only to the environment but also to the actual well-being of its citizens.

Further, the "how" questions are largely determined by technological capabilities, which are, in this era of globalization, often known and available virtually worldwide. An automobile is produced with essentially the same technology in China or in Europe, albeit with much lower wage rates in China. Computer-software technology in India is essentially the same (and quite interchangeable) as that in the United States. But, the countries in these examples are, nonetheless, seen as having very different economies. Why? Certainly, there are widely recognized political differences, but the key to the *economic* distinctions rests largely in the arena of *distribution.* For example, a command economy (one in which the government owns the land, resources, means of production, and therefore makes all economic decisions) and a market economy operate upon fundamentally different premises. To elaborate on this distinction, we must delve more deeply into the approach to distribution within a so-called "free-enterprise," capitalistic market economy.

Distribution: The All-Important, Ignored Element

Distribution is the forgotten branch in the discipline of economics. How is the output of the system allocated to the participants (i.e., citizens) of that system? How are the fruits of everyone's labors distributed? In a "zero-sum" world, what do you have that I do not have? We label it the *forgotten branch,* because, to devout, market-economy advocates, the prospect of having to make conscious decisions about these questions suggests the need for unnerving, normative judgments about value. Most economists cling to the hope that economics can be defended as a science, wherein only positive (mathematically derived), value-free conclusions and statements are allowed to exist—a human impossibility. Every question we humans ask is subjective and thus value laden. It cannot be otherwise. Therefore, every conclusion and statement is, by definition, subjective, whether mathematically derived or not.

Yet, how much more comfortable it is to contend loftily that "the impersonal market makes those decisions," and then proceed to expound in elaborate, mathematical detail on how this creates efficiency, rather than be seen making capricious, personal judgments about the distributional equity or fairness for other individuals, with such statements such as the following:

- "Aggregate human well-being would increase if we raised wages of the poor."
- "Increased taxes on the rich would be a good idea because the resultant decrease in their well-being would be less than the benefits of the resulting public expenditures."
- "A wilderness area ought to be saved because it's good for the human psyche, and biodiversity is valuable in and of itself."

The impersonality of the market alleviates the need for taking a position on such obviously important political issues, although many economists will, with only slight embarrassment, contend that they wear two hats. While they might be willing to make some such statements as a *citizen*, they are prevented from doing so as an *economist*. If this is true, it is a terrible indictment of their chosen discipline.

Economic Methodology Thwarts Redistribution

In this section, we may appear to go over some ground that is already covered earlier in the book. We view this as desirable, however, because here in our coverage of distribution is the proper place to emphasize the point that these types of questions constitute the lion's share of all controversy in social issues involving economic factors—whether that controversy stems from economic behavior in practice or from economic theory.

As noted earlier, a free-enterprise, capitalistic economy attempts to answer all three of the basic questions (What, How, and For Whom) through the mechanism of the market, supposedly in an impersonal and value-free manner with the aid of a "benign competitiveness." Early economic thinkers made considerable use of the word *subsistence*, which we defined economically as meeting bare human necessities. It was as though the grand purpose of an economy was to free humankind from the Dark Ages or even the cave people from the status of hunters-gatherers. Gradually, however, as the Industrial Revolution unfolded and technological change became standard fare, the success of an economic system came to be measured by the degree to which it could elevate its participants above the perceived subsistence level.

The "materialistic hierarchy" goes something like this: Primitive is bad. Developed (i.e., industrialized) is better. Rich is best. Can an almost religious fervor for economic growth be far behind? Today, of course, most economies are able to produce enough to sustain their citizens at something above the perceived level of pure subsistence. Supposedly, therefore, the question of current ability to produce enough is not the major economic challenge;

rather the important question becomes a distributional issue: What happens when there is a surplus, and who gets it?

Revisiting the Notion of Surplus

An intriguing way to categorize and assess any economic system is the manner in which it handles surplus. Surplus can be defined as production (and, of course, subsequent use or consumption) above the level of perceived subsistence. Moving beyond subsistence therefore involves the creation of surplus. Hence, when an economy gains the ability to generate products and incomes in excess of the necessities perceived as subsistence, that excess must somehow be distributed. Virtually every economic conflict throughout history can be fundamentally interpreted as fights over surplus.

Pursuing this point briefly can demystify the formal disciplinary attitude toward the field of economic development, including the methodological assumption that economics tacitly seems to make, namely that development is *always* positive (read mathematically derived and thus objective). Producing for subsistence at its most elementary level involves simply producing enough to satisfy the basic necessities of food, water, shelter, and space, which explains why an agrarian economy is often seen as "underdeveloped." Industrial production is therefore above the level of subsistence by definition, because it is seen as producing surplus. Thus, in the broad scheme of "more is always better, richer is always better," industrial production is thought to be at a higher level of development than is agricultural production.

More than any other factor, the method of distribution characterizes an economic system. Some broad examples will underscore this vitally important point. Early civilizations, such as the Egyptians and Romans, employed slaves to create surplus, maintained the slaves at the level of subsistence, and kept the surplus they produced to support the upper classes of Egyptian and Roman citizens in lavish lifestyles.

In Europe, on the other hand, little surplus existed during the Middle Ages. Due to Europe's complicated feudal system and the marked influence of religious thought, what surplus did exist found its way to the Church and to the ruling monarch and nobility, who were seen as earthly representatives of the Kingdom of God. The serfs were effectively the slaves of the Middle Ages.

Then, in the mid-18th century, the British Industrial Revolution substituted machines for human labor, which meant substituting the power of fossil fuel and steam for a slave economy, which used mainly wood and solar power. This all-important transformation of an economy from a primary source of energy, which is a renewable *flow* (solar energy), to one that is dependent on a secondary source of energy (fossil fuel, a nonrenewable resource or *stock* subject to depletion only) was accompanied by the ascendancy of a new, important actor in the economic drama—the *capitalist.*

The existence of *capital,* as the machines came to be called, naturally gave birth to the capitalist—the person owning the machines—and, of course, to *capitalism.* The latter term simply put an appropriate label on a system of production in which the components were in private ownership. Now, due to the harnessing of these seemingly wondrous new technologies, it was possible to create surplus at a pace that was historically unimaginable.

As might be expected, this increased tensions surrounding questions of how the surplus was to be distributed—questions that would not have surfaced if virtually everyone had been at the level of subsistence. But the capitalist was clearly above that level. For better or worse, conventional economic doctrine concluded that the capitalist should get to keep and use most of the surplus. After all, he (and it was virtually always a "he") owned the machines responsible for the dramatic increases in both production and the ever-widening gap between the requirements of subsistence and the availability of actual excess.

The story does not end here, however. Tracing the parallel development of this newly dominant economic system and the intellectual thought supporting it, we observe the capitalist keeping most of the surplus by the 19th century. This circumstance lands us squarely in the middle of the classical economics of Adam Smith, David Ricardo, Thomas Malthus, and Karl Marx.

Most of these thinkers celebrated the wonders of industrialization, specialization, and trade. However, the pollution and squalor of the working conditions in the early European factories were apparent to Karl Marx, the most heretical member of the classical economic school. Marx, as did all his less radical compatriots, accepted the labor theory of value, which contends that human labor is the basic, underlying source of all productive capability. Nevertheless, because he observed and reacted to the exploitation of workers, his view of the distribution of surplus differed from the other, more mainstream thinkers—most of whom are not only revered but also still quoted when it serves the particular vested interests.

Marx, however, is studiously ignored due to his unpopular politics, even though his economic analysis is clearly more accurate than that of his Classical School colleagues. The globalized corporate world of today, for instance, was predicted almost exactly by Marx, and bears little resemblance to the perfectly competitive and benign market economy of Smith and Ricardo. On the other hand, Malthus, with his gloomy premonitions of impending scarcity, may turn out to be the real oracle.

According to Marx, workers were customarily paid just enough to meet the required standard of subsistence, despite the fact that they (in conjunction with the emergent machines) were fundamentally responsible for the dramatic increases in the output of commodities. To Marx, therefore, the capitalists were unfairly expropriating the surplus, and the resultant pattern of *distribution* was unfair.

But economic methodology has erected formidable bulwarks against *normative* arguments (those based on value), which question the fairness of distributional practices and their results. For instance, arguments in defense of continual economic growth, as the ultimate answer to questions of distribution, contend that inequality is a necessary effect (the other side of the coin, if you will) of the accumulations of capital required for the large-investment projects deemed necessary to keep the economy moving. In their most virulent form, these arguments contend that concentrations of wealth (read "inequality") demonstrate to the worker, the underclass, or those just starting out, that they, too, can reap the benefits of the "land of opportunity" if only they are diligent and hard working. Recitation of this fairy tale could continue *ad infinitum*.

Although the fairy tale may have been more-or-less true for some people throughout the economic history of the United States, it has not been true for many others, no matter how hard they have worked. An old adage helps to explain this inequity: *It takes money to make money*. Such amounts of money are seldom available to the average person, no matter how hard that person works. With the foregoing in mind, it is helpful to review briefly the conventional wisdom of mainstream market economics on the issue of distribution.

Every economic system employs all sorts of resources or components of production. Here, however, we are focused on the human component, because the primary issue with respect to the distribution of income is the equity with which employees are treated in terms of monetary compensation for their labor. The question thus becomes one of how the products and services generated by an economy (usually symbolized by gross domestic product, or GDP) are supposedly reallocated or distributed to the participants in that economy, most of whom contributed in some way to its creation and maintenance.

The theory of a free-enterprise market economy is crystal clear on this point: The valuable products of an economic system are reallocated to the components of production (be they people or machines) in the proportion to which either the people or the machines contributed to the making of the products. Pay no attention to the fact that this proportion is difficult to determine empirically. In other words, whoever contributes to the production of the gross national product (GNP) theoretically gets to consume it in like measure.

Despite the superficially workable logic of this "benefits received" viewpoint, it is fraught with perils when examined from an overall philosophical perspective. Four points, listed without additional comment, define and highlight the structural underpinnings of the distribution theory in market capitalism:

- The "market" makes decisions impersonally.
- If you do not produce something, you do not get to consume.

- Some inequality exists because of innately unequal endowments of talent, skills, and training among people as workers.
- Benign competition exists at all levels of the economic process.

The results are Darwinian in that the most fit (most adaptable) survive; in addition to which, they are supposedly value free. Observably, some people in a mass economy fall through the cracks and either cannot work (are disabled or handicapped) or cannot find a job (are unemployed). Although it is hoped that unemployment is temporary, it is, nevertheless, tacitly assumed that the unemployed have little training, few skills, and would thus have low increments of productivity. Although capitalistic thinking reluctantly acknowledges that allowances must be made for these "problems" (i.e., disabled, handicapped, and unemployed people), much controversy has always existed, from the poorhouses of Charles Dickens to unemployment compensation of today, over questions concerning the nature of the support they deserve and how much. Despite its problem areas, the capitalistic approach to wage determination, and thus of surplus allocation, is a process much revered in the literature of free enterprise and market economics.

The sum total of all these conditions amounts to that which economists refer to as *efficiency* in the overall allocation of resources. This state of efficiency supposedly ensures that the resources available to a particular economy are employed in creating the maximum possible amount of human welfare within that society or economic system. Although this intellectual construct is a product of capitalistic, market-based economics, a similar condition of efficiency is supposedly attainable for any economic system, be it large or small, rich or poor, complex or simple. It refers to an economy's use of its resources (employment, harvesting, mining, depletion) in the production of goods and services for distribution and ultimate consumption to satisfy human desires as much as possible.

The optimistic conclusion is that the market contains enough magic to automatically handle virtually every question dealing with distribution, and that the resultant inequalities are explainable and acceptable. But, of course, this says nothing about equity or fairness of the resultant patterns of distribution and wealth. The fairness of distributional patterns and endowments of wealth is supposedly a political question, not an economic one. Thus, by its own internal logic, economics does not concern itself with fairness.

Inequality and Economic Realities

Despite this utopian notion, some simple examples will assist in understanding the real effects of income distribution on an economy. Assume, for example, that two people form a rudimentary economy as sole producers and consumers. There is $100 of income between them, which at first blush we assume is evenly split, with each having a $50 per year income. What will be produced? Without specifying details, their demand for goods will be roughly similar, even allowing some differences for their individual tastes in choosing items beyond their basic necessities. After all, their purchasing power is identical. Therefore, as producers, even with some differences in productive skills and specialization, they will each be engaged in producing what they both need and want.

As a result, the important conclusion is that the existence of similar purchasing power (not necessarily similar tastes or similar work skills) on the part of participants in an economy will create the realization of interdependency between and among those producers and consumers. It will also create the necessary awareness, on both of their parts, of their *dual role* as both a producer and a consumer in this simple society.

Now, let us adjust the assumptions somewhat by asserting that one person has an income of $70, while the other has an income of $30. What will be produced in this economy? The answer stems from the obvious fact that the majority of the purchasing power—70 percent to be exact—rests with one of the two citizens of this system. Therefore, if everything is sold, 70 percent of the total output must cater to the desires of the wealthier person. In turn, the person with only 30 percent of the income will likely find himself or herself producing goods with the wants of the other person in mind—and perhaps struggling a bit more to meet his or her own needs.

The situation will be even more striking if we assume one person has 90 percent of the income and the other has 10 percent. Production will be almost totally oriented to the demands of the wealthier individual, perhaps even including people working directly for them as maids, servants, security guards, and the like. The choices of what to produce in the economy will shift dramatically toward services and luxuries, and away from the production of basic necessities, despite their increasing unavailability at the lower end of the income distribution. Incidentally, as we all know, the rich-to-poor ratio in the United States is *many times* worse than the nine-to-one ratio of this example—but read on.

There are two vitally important conclusions to be drawn from the various scenarios of this bare-bones example. The first is an economic observation. Namely, it is immaterial how affluent a system might be or how far above subsistence the typical worker/consumer might be when the average income is calculated. If the distribution of income becomes increasingly disparate, the least well off in the system are sooner or later driven to subsistence and poverty. This is true because the wealthiest exist in the lap of luxury, and the economy will continue to serve their needs because they have all the purchasing power.

This is a recipe for what we euphemistically refer to as a "Banana Belt Republic," which is a situation in which the majority of a population exist in abject poverty, while a very few exist in a level of wealth rivaling that of the richest people in the world. We in the United States scornfully reject such systems, but data on the distribution of income reveal that we are rapidly approaching just such a structure. The top 1 percent of our national population earns about 18 percent of the income (and rising), which is more than the combined earnings of the bottom 50 percent of the American citizenry. In terms of accumulated wealth, the top 1 percent of our population owns 34 percent, which is more than is owned by the bottom 90 percent of the citizenry.

In fact, recent figures show that the top 1 percent own 84 percent of all stock in the publicly held stock markets. With each upward tick of the Dow Jones average, even though we are all conditioned to rejoice, the real effect is that inequality gets worse. However, the most bothersome point may well be that the only other time in the nation's history that inequality approached these levels was 1929.

The second point to be made is political. When each participant has 50 percent of the income, the influence each person might perceive they have in any government action is necessarily similar. One person/one vote would be reasonable, and the political system would tend toward democracy. But, as the income disparity increases, wealthy people and poorer people perceive the benefits of government actions differently. Consequently, politics becomes more contentious.

In fact, as we are actually observing, wealthier people rightly perceive they can afford to purchase for themselves many of the traditional government services, such as education, health care, and even protection from exposure to environmental damage. Therefore, they may favor a minimalist approach to government, even while less affluent citizens would like to see a broader social safety net available for everyone.

In short, uneven income distribution leads to strident political conflict. Despite the fact that the wealthy are few in number, they assert control of the political process because they have the money and power to do so. One person/one vote is thus a threat to their continued control, a threat that leads not only away from democracy but also toward a plutocracy or even fascism or a dictatorship. Sound familiar?

If the problem of economic disparity is so obvious, why have we not done more about it? A strong notion of a middle-class embodies a cultural image of the American Dream—that everyone can become well-to-do. The overly simplistic, hypothetical example assumes, in effect, a zero-sum game. It is clear that wealth for the rich impoverishes the poor. Our devotion to continual growth has clouded that vision. If the pie can be made bigger, there is less of a case for resenting, or politically reacting against, anyone who currently has a disproportionately large piece of that pie. Given the complicity of mass media, we are even encouraged to admire, honor, and obey such people. The likes of *Fortune* magazine, with its annual chronicling of the "horse race of the rich," see to that.

The lesson taught by the Era of Limits is that making the pie bigger becomes increasingly difficult—even impossible. Moreover, laissez-faire capitalism inexorably worsens the distribution of income and wealth, which makes any restoration of social justice more difficult to accomplish. Consequently, the planetary squeeze has begun, and the zero-sum nature of our distributional choices is daily becoming more apparent, including, among other things, the steady disappearance of the middle-class. The first step in this zero-sum game is informed denial; the second is political conflict—both domestically and internationally. And we continue to observe an increase in both.

Equity and Social Justice—The Key to Real Sustainability

On the surface, we may appear to have arrived at the conclusion that for an economic system to be sustainable, decisions about the distribution of income must be made *consciously* by human choice, rather than unconsciously by the market. We are not sure this is true. In fact, we cannot imagine a central board, agency, or commission—in the manner of a centralized, planned economy—being able to do such a job effectively. Further, we do not have specific recommendations with respect to how wealth should ideally be apportioned in an economy, with appropriate salaries, income levels, and so forth. That would be folly. Rather, we are certain of one thing: An ethic of compassion, justice, social-environmental sustainability, and the recognition of planetary limits must be totally inculcated into whatever process exists in communities, states, and regions for the determination of issues related to levels of salary and income per individual. Clearly, these are *value judgments.*

If this admittedly difficult task were accomplished, the results would propel society in the direction of a sustainable, monetary distribution. Although we continue to assert there is no such thing as a *free* market, there can and must be such a thing as a *fair* market—one inculcated with the values identified above. If such an economic system was created within a local or regional framework, then we can imagine a constructive role for a labor market in creating a wage/price structure with much healthier patterns of distributional equity than those we currently observe.

To live—or work—justly, one has to know that the rules by which one lives are morally just, why they are just, and that one can live by them safely. Is any dominant societal notion of justice really fair in the moral sense, or is it merely something that is socially and legally acceptable at this moment in history, when money and political power dominate spirituality and compassion?

The *ideals* of justice, which so desperately need to be taught to those who will become parents and leaders, seem, in this consumerist-dominated culture, to be less and less apparent in homes and schools and, thus, in society at large. Consider, for example, the brief period in the history of Zimbabwe, in southern Africa, as it played out in the year 2000. We choose this graphic scenario because it represents the repetitive, economic competition for nature's resources based on a sense of moral superiority; economic competition; and raw, self-centered political power.

Harare, Zimbabwe: "Zimbabwe's government, facing a parliamentary general election in June [24–25, 2000], has enacted a new law to allow Robert Mugabe, the President, to seize up to 841 white-owned farms without paying compensation for the land." President Mugabe decreed that, while whites may live in Zimbabwe, they would never have a voice equal to that of blacks.

"The whites can be citizens in our country, or residents," said Mugabe in a campaign speech one week prior to the parliamentary elections, "but not our cousins." According to Mugabe, who repeatedly attacked the small, white minority through vitriolic language, "They are the greatest racists in the world. Now, the British are saying that they [the blacks] are squatting on white man's land. Where is black man's land in Europe? Zimbabwe is a black man's land, and black men will determine who gets it." He went on to say that "Our present state of mind is that you [the white farmers] are now enemies because you really behaved as enemies of Zimbabwe." The British had lost Rhodesia, now Zimbabwe, as a colony more than 20 years earlier.[1]

> Mugabe also expressed sympathy with the mobs of blacks who illegally occupied more than 1,400 white-owned farms, because one-third of the fertile land was still owned by about 4,000 whites. He called the occupations a justified and legitimate protest against the unfair legacy of land distribution left over from the days when Zimbabwe was a British colony.

Violence erupted. Black squatters, who would undoubtedly dub themselves "freedom fighters," rampaged through a farm worker's village, kicking in doors, smashing windows, and burning down about 30 homes as the dismayed workers watched. Two farmers were killed and at least a half-dozen beaten. They also beat a white farmer unconscious, fractured his skull, broke his arms, and left him to die. Chenjerai Hunzvi, the leader of the black squatters, said this latest killing was not worthy of comment.

A snippet in *The Wall Street Journal*, dated November 13, 2001, stated that: Zimbabwe banned 1,000 white farmers from cultivating their fields and gave them three months to vacate [their] properties. The presidential decree means [that] legal challenges won't halt settlement of black families on the land.[1] Mugabe's opponents contended that his 20-year, authoritarian regime had resorted to race-baiting and fanning the flames of these simmering grudges over the distribution of land in order to bolster his flagging political popularity. As a result, the invasions of white farms, combined with the orders forbidding white farmers to work on their property, had reached the point by August 2002 that forced the production of grain to be severely cut. In essence, Mugabe's program of land redistribution, based on the perception of past inequities, had forged ahead, and six million Zimbabweans, roughly half of the nation's people, faced potential starvation.[1]

Our overall observation is that the distributional conflict, largely over pride and power, resulted in the material inadequacy of a livelihood for millions. The example speaks both to distributional inequity as well as to the waste of natural resources engendered by conflict.

In contrast to the Zimbabwe tragedy, 75 soldiers from South Africa emerged as unlikely heroes, while the floodwaters, which had ravaged neighboring Mozambique during February and March of 2000, began to recede. Heroes? Yes, heroes because they almost single-handedly rescued nearly 15,000 people who were marooned on rooftops and in treetops and were clinging to utility poles, as the rest of the world watched, wondered what to do, and decided whether to act.

"We owe the South Africans a great deal of gratitude,"[1] said Silvana Langa, director of Mozambique's disaster relief agency, an irony lost on no one old enough to remember when the South African white minority not only ruled South Africa through apartheid but also was at war with Mozambique. In fact, it is likely that South Africa had once been Mozambique's worst enemy, because it tried to topple the nation's socialist regime by supporting the insurgents. But much had changed in South Africa during the decade of the 1990s.

For example, the white soldiers and the black guerrilla fighters, once sworn enemies, formed a single army committed to rescuing Mozambicans from a watery grave. "I don't think any of us thought about the irony of the past," said Brigadier General John Church, a veteran of 33 years in the South African Air Force. "We weren't thinking about who was white, and who was black, and history. We were just trying to save as many people as we could. . . . We just saw it as Africans helping Africans."[2] The message here is consistent with the Zimbabwe story but has the opposite outcome: Cooperation and equity create material adequacy.

With the foregoing in mind, we reiterate that promoting an equitable income distribution is a *value-based* concept, where the agreed-upon premises consistent with social-environmental sustainability are unapologetically allowed to dominate the so-called *value-free* impersonality of the overall global marketplace. As we have said, far from being value free, the global marketplace (both for products and inputs) is actually inculcated with strong value judgments made, in fact, by the world's economic elite, and, as such, has profound, equity-destructive distributional characteristics. The resultant inequality pulls the world in powerful directions *away* from sustainability (reminiscent of the blatant, social-economic inequity in the foregoing story of Zimbabwe), especially as it seeks to rely on growth as the answer to poverty and the uneven distribution of wealth.

The important conclusions for this chapter on distribution—a topic that necessarily injects normative decision making into a highly resistant and suspicious discipline—are twofold. First, given the impossibility of permanent growth within our biosphere, we are forced, by our ecological crisis, to find a better solution. Second, poor distribution means that a given resource base for an economy, or for the world as a whole, is more likely to be deemed inadequate to meet human necessities by those in control who coincidentally have all they want. On the other hand, that same resource base is more likely to approach adequacy if given a better distribution of resources (monetary and otherwise), because equity breeds adequacy.

From the viewpoint of macroeconomic policy, poor distribution makes all economic problems harder to resolve, and better distribution makes them easier. Distribution is at the core of the most pressing social, economic, and ecological challenges of our day.

Once again, we are compelled to return to central tenets directing this work. The overriding conclusion is one based on interdependency: *A fair, just, and equitable distributional system promotes an economy that approaches a condition of social-environmental sustainability; conversely, an economy that approaches a condition of social-environmental sustainability leads to a fair, just, and equitable system of distribution.* It is a self-reinforcing feedback loop. Therefore, we cannot just solve pieces of the puzzle in a symptomatic sense; we must address the whole picture in a systemic sense. As a final note of optimism, evolving to a systemic mindset might allow all the pieces to fall into place and may actually be an easier task to accomplish than the alternative, which is the symptomatic, piecemeal reformation of a currently pathological economic system. Next, we conclude Section II by looking more closely at the macroeconomic system.

Endnotes

1. The following discussion of attacks on white farmers in Zimbabwe is based on Zimbabwe Enacts New Law To Seize White Farmers' Land. 2000. *National Post*, Toronto, Canada. May 25; Paul Salopex. 2000. Attacks on Farms Continue in Zimbabwe. *Chicago Tribune*. In: *Corvallis Gazette-Times*, Corvallis, OR. April 19; Angus Shaw. 2000. Another White-Owned Farm Attacked in Zimbabwe. The Associated Press. In: *Corvallis Gazette-Times*, Corvallis, OR. April 21; Susanna Loof. 2000. Attack on White Farmer Adds to Toll in Zimbabwe Uprising. The Associated Press. In: *The Sacramento Bee*, Sacramento, CA. May 8; Anonymous. 2000. Call to Round Up, Expel Whites in Zimbabwe. *San Francisco Chronicle*, San Francisco, CA. May 8; Anonymous. 2000. Mobs Rampage through Zimbabwe. 2000. *USA Today*. May 9; Anonymous. 2000. Mugabe: Whites Must Realize That Zimbabwe Is for Blacks. The Associated Press. In: Albany (OR) *Democrat-Herald*, Corvallis (OR) *Gazette-Times*. June 18; Laurie Goering. 2002. White Farmers Forced from Land. *Chicago Tribune*. In: *Corvallis Gazette-Times*, Corvallis, OR. August 15.
2. Jon Jeter. 2000. South African Troops Emerge as Unlikely Heroes in Flood. *The Washington Post*. In: Albany (OR) *Democrat-Herald*, Corvallis (OR) *Gazette-Times*. March 12.

8

Macroeconomics—Is It Still Helpful in an Age of Scarcity?

Macroeconomics is defined as the study of a whole economy or, in other words, a relatively closed economic system, as opposed to its individual parts. A study of the components or building blocks of the overall economy (such as individual consumers, businesses, and markets) is the purview of microeconomics, the core of which is the analysis of supply and demand. Microeconomics is essentially what we have done to date in this examination of economic methodology; but it is time to look at the big picture.

Origins of Macroeconomics

As a major branch of the overall discipline of economics, macroeconomics is a child of the Great Depression and was therefore born out of necessity. Prior to the economic collapse of the 1930s—ostensibly of the entire capitalist world—the dominant mode of economic thought in the United States can accurately be termed *market capitalism,* although perhaps a much-simplified version of what that term would suggest in today's globalized economy. Markets were assumed to be self-correcting. There was no such thing as a shortage or surplus, but merely a price that was too low or too high. And the pricing mechanism would quickly rectify any supposed imbalance. The Great Depression called these comforting assumptions into question.

In short, virtually all economics was microeconomics, and the minimalist, self-correcting nature of the methodology fit the times. This ideology accompanied the predatory growth of the capital goods industries during the last half of the 19th century—often termed the Robber Baron era—and ushered in the beginnings of the consumer society, culminating in the euphoric Roaring Twenties. Clearly, the role of government in the economy was minimal during this period, with the notable exception of the antitrust efforts through the Sherman Act of 1890 and the establishment of the Federal Reserve System (the Fed) in 1913.

The Sherman Act addressed the perceived egregious monopoly and excessive concentration of economic power that constituted the Robber Baron era, while the Fed established mechanisms for regulating the money supply in order to deal with the observed tendency of capitalism to experience periodic financial panics. But, in general, laissez-faire was the order of the day. It should be noted that antagonism toward activist government involvement in the economy did not resemble the sophisticated ideological arguments of today, but rather was something that, in a democratic society, had never been tried or even really considered.

That all changed following the 1929 crash of the stock market, which ushered in the Great Depression of the 1930s. However, in the absence of any experience with analytical tools that might suggest appropriate actions for the public sector, governments were at a loss as to what to do. One possible set of answers came from a British economist and Chancellor of the Exchequer, John Maynard Keynes. To professional economists, his landmark 1936 treatise, *The General Theory of Employment, Interest and Money,* marks not only the initial public exposure to a promising road map out of the economic morass but also the beginning of macroeconomic theory as a subfield in economics.

Keynes sought to develop an approach to economic policy that could effectively address the worldwide crisis of the Great Depression. Broadly speaking, he sought to save capitalism, which during the 1930s was assumed by many to be a failing system. This is an ironic point, because subsequent conservative critics of Keynes have often accused his teachings and his followers of destroying, or at least completely impairing, the supposedly free-enterprise capitalistic system. Consequently, he laid out a framework for government action that has come to be known as compensatory financing.

With historical perspective on our emergence from the dismal 1930s and the several decades of apparent middle-class prosperity that followed, we must ask the question: Did Keynes succeed, or did the elevated levels of public expenditure forced by World War II have an unintended, but beneficial, economic effect of ending the Great Depression? This question may not be answerable, and that is not necessary. However, given that his framework is generally judged to have worked, an important question for us now is as follows: Can we once again use the framework of macroeconomics to pull ourselves out of the economic quagmire that became apparent in 2008, which many are calling the Great Recession? Addressing this question is a major task of this chapter.

Basic Macroeconomic Worldview

Keynes developed a framework whereby direct and conscious intervention in the economy of a nation could be undertaken for the purpose of controlling or improving the performance of that economy. The direct tools are such things as taxation, government spending (fiscal policy) and regulation of the money supply, interest rates, and terms of credit (monetary policy). The current bailout efforts and stimulus packages are cases in point.

Immediately, this made macroeconomics controversial, because there was little or no historic precedent for activist government involvement in the private economy. Both the established, behavioral norms of the laissez-faire ideology and the notion of a self-regulating (microeconomic-dominated) economy were powerful forces in support of business as usual.

As a personal note, when I (RB) began learning conventional economics in the late 1950s, I was told these ideas of Keynes were still viewed by many in the field as "left-wing socialist." Nevertheless, as these ideas, which prefigured macroeconomics, worked their way into the mainstream of economic thought, they did so under the banner of the "Harvard School," championed by the likes of Paul Samuelson and John Kenneth Galbraith. But, established ways always create powerful resistance to change. Thus, it comes as no surprise that formal opposition to this activist approach to economic policy flourished within the more conservative "Chicago School," represented primarily by Milton Friedman. Even though colleges with an economic curriculum have long taught both microeconomics *and* macroeconomics, most major universities have gradually developed a personality and a reputation that fall generally in line with one of these two schools of thought.

These developments are more than mere academic trivia of interest only to those in the profession. They have dramatically affected our lives and our politics. When Ronald Reagan became president in 1981, for instance, Milton Friedman became an influential advisor. It is no accident that the *neo-conservative* movement promotes thinly disguised attempts to remove any remaining vestiges of the New Deal, including weakening or privatizing Social Security, opposing welfare and government public-works projects, as well as criticizing such programs as Medicare. (A neo-conservative is part of a U.S.-based political movement rooted in the aliberal, anticommunist movement of the Cold War, and is a backlash to the social liberation movements of the 1960s and 1970s. These liberals, who drifted toward the conservative right, are neoconservatives, from the Greek *neos* or "new." They not only favor an aggressive, unilateral use of economic and military power as foreign policy but also generally believe that the social elite protects democracy from mob rule.)[1]

The ideas of Milton Friedman are even cited as the philosophy underlying the development efforts in many foreign countries (e.g., Chile). The implementation of Friedman's philosophy has led to international debt, which usually originated through the corporate-dominated World Bank or International Monetary Fund. In turn, these debts have led to the impoverishment of indigenous peoples and the privatization and corporate takeover of valuable natural resources through bogus patents in numerous, nonindustrialized countries.[2] Certainly, the influences of globalization are increasingly pervasive since the early 1980s as a result of the prominent influence of this line of thinking. (See, in particular, Naomi Klein's *The Shock Doctrine:*

The Rise of Disaster Capitalism,[2] and *Confessions of an Economic Hit Man*[3] by John Perkins.)

To be sure, the boom and bust tendencies of capitalism have been well chronicled. The theory of the business cycle goes back at least to the writings of William Stanley Jevons in the mid 19th century. Thoughtful analysts realized that market economies were capable of erratic behavior, but we had always managed to muddle through with the perception that the benefits of the boom periods would outweigh the pain of recessions and the periodic financial crises.

The Keynesian Dilemma—Unemployment or Inflation?

So what does compensatory financing look like or amount to? In recognition of capitalistic instability, compensatory financing seeks to smooth out the extremes (the peaks and valleys) of the business cycle. Having said that, we must identify the negative characteristics of these peaks and valleys and determine why it is desirable to avoid or minimize them.

In the first place, the cycle can be imagined merely as a line going up and down on a graph, which traces the level of all business activity over time. For a particular economy, it is something akin to the total economic output of a given economy, often termed the gross domestic product (GDP). Viewing it from the other side of the coin, it is like the collective, national income of the people involved in this economy. More importantly, it symbolizes the essence of prices, wages, resource needs, employment, and the total volume of goods and services available to be used in meeting human necessities *and* desires of that particular society.

In that context, a *peak* (boom period) symbolizes high levels of employment, perhaps even labor shortages, and potential inflation in the form of upward pressure on both wages and prices. There may also be shortages or bottlenecks of necessary goods or resources. In the vernacular of economists, the economy is *overheating*.

In the valleys, or *troughs* (bust period) of the cycle, the problem is unemployment, idle resources, underproduction, and a general failure of the economy to support the basic material requirements of its citizens. There may even be wage/price *deflation*, which can have consequences as serious as inflation. In addition, even the rapid fluctuations are deemed to be a problem in that they undermine certainty and business expectations and thus negatively affect people's willingness to invest and thereby promote economic growth.

Although there are clearly several possible negative impacts stemming from fluctuations in the business cycle, the major problems within a capitalist economy can be simply summarized as *inflation* in the peaks and *unemployment* in the troughs. Macroeconomic theory and policy begin with the premise that capitalism innately tends to be somewhat unstable and is always on the horns of a dilemma between unemployment and inflation. Therefore, in their simplest forms, the control mechanisms of monetary and fiscal policy are built around bolstering employment in the troughs, fighting inflation at the peaks, and thus promoting overall stability of the economy.

Controlling the Economy

And how do we propose to do that? At the very outset, the policy tools recognize that the government, beyond being the place where we collectively make and enforce the agreed-upon rules for the society, is a significant demander and supplier of goods and services in the economy—hence, the term *public sector*. In other words, the government is recognized as an economic entity with considerable power.

Right here, an astute reader can sense controversy. Many times, we have all heard, from a conservative or libertarian mindset, that the role of government should be absolutely minimal. Its role should be to operate the courts, enforce private contracts, and provide for national defense. The latter is perhaps the only area (i.e., military hardware and personnel) in which the government should legitimately be allowed to be a significant employer or purchaser of goods and services. In its more extreme versions, such a worldview implies little or no support for public education, social security, Medicare, and certainly public welfare programs.

Therefore, just the thought of engaging in fiscal and monetary policy invokes skepticism and outright ideological opposition, even before the actual fact of implementing specific measures emerges from the starting gate. Reasons for such opposition become more apparent as we examine the nature of those tools in greater detail.

Monetary policy involves the buying and selling of government bonds by the Fed, and the increase or decrease of the discount rate, which is a key interest rate at which banks borrow, and to which mortgage rates, bond rates, and a host of other key interest rates are related. Further, the Fed establishes the reserve requirement, stipulating the fraction of assets a given commercial bank needs to keep on deposit with the Fed in order to legally make loans to businesses and individuals. Such loans create checking accounts, or demand deposits, which are the primary component of the money supply. Through these tools, the Fed exercises considerable control over the financial sector of our economy, including interest rates, other terms of credit, the money supply, and, indirectly, the volume of business activity to the degree that it is financed by borrowing.

Fiscal policy deals with the level of taxation and government spending and is determined by government decisions, made through the auspice of the Treasury Department. Whereas fiscal policy is the bailiwick of the government, monetary policy is the purview of the Fed, acting as an independent, quasi-public entity, which is actually owned by its member banks. Whereas fiscal actions are necessary to operate the business of government (meet salaries, make committed expenditures, etc.) and are only secondarily an instrument of economic policy, monetary policy actions by the Fed are done strictly for purposes of attempting to control or direct the economy.

According to Keynes, the broad-brush prescription for dealing with an overheating "boom" phase of the cycle is for government to tax at a surplus greater than public-sector spending (i.e., reduce the government debt) and for the Fed to engage in restrictive monetary policy that indirectly restricts the money supply—the direct control of which is actually in the hands of private member banks as they make loans and extend credit to businesses and individuals. Such actions will serve to dampen the level of economic activity, control the financial sector, and consequently hold down inflation.

In a recession or trough, the prescription is the opposite. The Fed should attempt to lower interest rates, ease terms of credit, and take actions that would stimulate monetary expansion and borrowing for purposes of supporting higher levels of business activity. In such an environment, inflation is presumably less of a problem than is unemployment and the need to create jobs. Therefore, the fiscal policy called for in recession is for government to spend more than it takes in through taxes (increase the public debt), which injects new demand into the ailing economy.

This process of dealing with recession should, according to Keynes and his followers, be called "pump-priming," because the perceived problem is that the private-sector economy has temporarily inadequate levels of demand for goods and services. Accordingly, the government would fill the gap through spending on public works, which would stimulate the private sector to resume investing, spending, and creating jobs, after which the government could retire the debt as the economy pulled out of the trough (and headed toward another boom period?).

The entire process is termed *compensatory financing*, or *countercyclical economic policy*, because the role of public policy can be seen as a monetary mechanism with which to stabilize or counter the direction that an innately volatile and unstable (though hopefully vibrant), private, capitalistic market economy will head if unchecked. An astute reader should have no trouble perceiving that the operational questions, loaded with huge political implications, become the following: How much government control is too much? and Who in the government knows how to do the right thing at the right time?

In theory, although some level of federal debt exists, it should not increase permanently if the private-sector economy (despite understandable, cyclic fluctuations) retains its overall long-term health. In practice, as our current economic malaise indicates, getting out of the trough of recession has historically proven more difficult than dampening the peaks. One way of putting this is the homely expression, "You can pull on a string, but you can't push." Interest rates, for instance, are currently near zero percent, but without adequate demand in the economy, even this has not stimulated new business investment. The blatantly obvious levels of rancorous controversy over bailouts, troubled asset relief programs (TARPs), and stimulus programs is testimony to the ongoing historic differences of opinion not only about the effectiveness but also about the ideal level of government intervention in our economy. Rarely in our history, however, has this age-old debate so polarized our national politics.

Revisiting the Capitalist Scenario

At this juncture, it is appropriate to reintroduce some concepts raised in previous chapters, wherein we contended that further exploration of those issues should await coverage in the macroeconomic section. That time has come. During earlier coverage of production and productivity, the point was made that economic theory is actually designed to address the productivity of all inputs and not merely that of labor. That point was reiterated in the previous chapter on distribution.

It is understandable that worker productivity, or the contribution of the human input, received the lion's share of the attention at those junctures, because it is, after all, the well-being of people that should be the ultimate focus of any economic system. But *macro*, as a term, connotes a comprehensive overall analysis—the big picture, if you will. Therefore, it is the time and place to seriously consider issues related to the role of other factors of production, such as land, capital, energy, and any of a number of various raw materials, such as water, wood, and minerals.

Clearly, any consideration of all these other nonhuman inputs shifts the focus back to the undisputable fact that the economy depends squarely on the natural environment. From the macroeconomic point of view, therefore, attention must, by definition, be on the mix of inputs and endowments of the world's resources, and not just on human labor. As we will see, this balanced systemic perspective allows insights into the role of labor, which are not apparent through a more targeted consideration of human productivity alone. But first, what are we able to say about all these other factors of production?

There is ironclad logic behind this approach. World production clearly depends on all resources that might go into the production of any (or all) product(s). As the Capitalist Scenario at the end of Chapter 5 contends, the production of any product will be first constrained by the most scarce resource, or economic input, whatever that might be. And, given the extreme worries, especially in the United States, about the unacceptably high levels of unemployment (just under 10% at this writing), the constraining variable on world output is guaranteed *not* to be labor. Unemployment means, by definition, that labor is a superabundant resource.

Therefore, assessment of the productive capability of any macroeconomy—and of the global economy—demands a balanced examination of the complete range of all other potentially useful economic resources. This means, given the point we have repeatedly emphasized, that all such resources are in some way extracted from the environment, and that accurate economic assessments can only be carried out in conjunction with an equally comprehensive environmental assessment. In fact, they must, for a sustainable future, become virtually the same analysis.

Two methodological points must be mentioned in conjunction with any robust and useful macroeconomic assessment of the long-term availability of climate, land, air, water, minerals, energy, and so on. These refer to the vital contributions of the biosphere, without which an economy of any kind would be impossible. The first question is one of *quantity*. Are the required resources permanently available in the desired amounts, or are they *depletable*? The second question is one of *quality*. Are resources available in the condition required, or are they *degradable*? We must ask where quantity ends and quality begins, and can we tell them apart?

To elaborate, land, minerals, and fossil-fuel energy are examples of resources subject to depletion. Air, water, soil, and climate are examples of degradable resources. The quality of air, water, and soil may be inadequate to support healthy human life or to be used in a production process. Thus, a resource "failure" for human societies can either mean that the resource is effectively gone (exhausted), or that it is degraded beyond the point of usefulness (polluted).

Further, these tests occasionally interact, and we can be lulled into thinking our major problem is one, when, in fact, it is the other. For instance, many analysts and policy makers, and the public at large, have long worried about the specter of running out of oil. Currently, many climate specialists contend that given the carbon saturation implications of burning fossil fuels, the environmental carrying capacity will be destroyed long before we ever burn all the known fossil fuel (or perhaps even just petroleum) reserves. In other words, we may think we have a depletion (quantity) problem when we actually have a degradation (quality) problem, which is guaranteed to dominate. If these predictions are true, this means that the environment will not even allow us to consume all that is available. What are the chances of taking the correct steps in promoting socioeconomic sustainability if we are not even responding to the real problems?

Little more needs to be said about the status of the world's resources that has not been acknowledged. Both depletion and degradation are threatened in many areas: oceans, fresh water, forests, clean air, plant and animal biodiversity, and the all-important characteristic of climate change. We do not need to go into resource-by-resource specifics here, and that is not our purpose. Rather, in order to properly conclude this section dealing with macroeconomics, we need only note that in assessing any one of them, individually or collectively, as an *economic input*, within the current growth-oriented worldview, *they are all more threatened than labor*. In other words, we have overpopulated the world (and the world of work), which makes it critical that *economic* solutions and *ecological* solutions are, in effect, identical. Anything else will fail. Symptomatic treatment of problems within the old paradigm will *not* work. The dominant worldview for even addressing the problems must change.

Finally, the Capitalist Scenario pointed out the tragedy perpetrated by seeking to support people through infusions of automation and reliance on capital, which can now be defined merely as the conversion potential of the natural environment into products. We built a system that relies intensely on increasingly scarce resources and, furthermore, one that discards, discredits, and underpays the element it was all intended to serve: people. In order to elicit final conclusions, we return to the starting point for this chapter—Keynesian macroeconomics.

Age of Scarcity Changes the Paradigm

As we chronicled, the Era of Limits can be said to have begun approximately in the late 1960s and 1970s. At the very least, this is when the specter of scarcities first received any level of general public attention. The world of the 1930s was a different world. The problems of economic failure were ascribed to inadequate use of the resources that we had. The difficulty has been described as idle capacity, not only human labor but also such resources as land, capital, energy, and so on. We were seen as not employing, or as underemploying, the resources that were available. The need was to get everything and everybody back to work.

Then, as now, there was widespread unemployment. And this familiar need to provide jobs and get people back to work can seduce the powers that be into thinking that we have the same problem with which we were confronted eight decades ago and, therefore, that the same types of solutions will suffice. Recall that the Keynesian prescription, which evolved into conventional macroeconomic dogma, begins with the assumption that capitalism careens between unemployment and inflation.

There is today a vital difference: the resources that were observably idle in the 1930s are now scarce, overworked, and deteriorating as we begin the 21st century. We still have unemployment, in part because we have so many more people, but the situation with the rest of the natural-resource base is exactly the opposite. Scarcity has replaced idleness. Availabilities of a host of necessary resources are in question. Once-pristine environmental resources are polluted and degraded. The macroeconomics, which may once have salvaged capitalism, will, in all likelihood, not be able to repeat the task. We say this because there can be little hope of constructive results from a tool based on assumptions to meet a time in the world that was the antithesis of the situation that actually exists in today's world.

A Growing Economy, a Planet in Peril

Conventionally, classic macroeconomic policy in the current high-unemployment morass would, at this point, sound a familiar refrain: Stimulate the economy and promote economic growth. Perpetual economic expansion is conventionally seen as the surest route to save us from the trap of stagnation and keep open the door to continual increases in the standard of living for each citizen as worker/consumer. But this Holy Grail of uninterrupted prosperity carries two prerequisites. First, there is need for a constantly increasing investment of capital (and, as we have seen, infusions of energy) to support the biophysical requirements imposed on the environment—both natural and social.

Second, technological innovation must support constant gains in worker productivity. In short, both the *quantity* and *quality* of capital must respond accordingly. And even should these conditions be possible, and all information about planetary limits dictate they are not, the ironies of the Capitalist Scenario tend to powerfully restrict, rather than expand, employment, and tend to drive wages down, not up.

Two enlightening side comments are appropriate at this juncture. First, this is very close to the underlying theory behind the economics (not the politics) of Karl Marx. The theory sets up and explains—virtually requires—the classic conflict between capital and labor (owner and worker, proletariat and bourgeoisie) that has characterized much of our economic history. Second, in the current world of economic globalization, this theory both supports and explains the controversial outsourcing of jobs, as capitalistic decision makers seek not only to increase capital relative to labor with the unerring assistance of new technologies but also to replace expensive workers directly and immediately with cheaper labor. Exploring these topics in any depth would require separate books.

Finally, there is no justification for picking on Keynes. He was a genius whose innovative mind precipitated an economic revolution, and he was correct for his times. The challenge is ours to realize that we live in a different world and that a different and compatible new worldview must emerge. First, however, some strictly economic comments are in order. Ironically, labor must be seen as the superabundant resource and must be used extensively with the scarce capital resources of the biosphere—including the environment. Favored processes must be labor extensive and capital saving—precisely the opposite of past capitalist ideology.

We offer a caution, as the Capitalist Scenario implies this will, in all likelihood, result in lower salaries to workers. After all, they are the superabundant resource. But we have already pushed the world past any distributional option for equitable "high salary/high consumption" by allowing the human population to dramatically exceed any reasonable planetary carrying capacity. This has precluded any alternative involving an average, worldwide standard of living that is anything close to the current U.S. or European levels. We have not obeyed nature's biophysical principles, and nature bats last. People, as a species, will need to learn to live with less, but our planetary home will survive.

This seemingly harsh set of ideas offers a platform for launching us into the concluding part of this book. Might the end of the consumer society be a happier world? We think that it can, provided we make adjustments in our thinking that might support appropriate changes in the dominant paradigm.

Endnotes

1. Neo-conservative. www.sourcewatch.org/index.php?title=Neo-conservative (accessed on December 11, 2010).
2. Naomi Klein. *The Shock Doctrine: The Rise of Disaster Capitalism*. Metropolitan Books, New York. 2007.
3. John Perkins. *Confessions of an Economic Hit Man*. Berrett-Koehler, San Francisco, CA. 2004.

Section III

Reconciliation and Looking to the Future

9

The Meaning of Social-Environmental Sustainability

Social-environmental sustainability is just another term for *intergenerational equity* and refers to the responsibility of the current generation to its own members, its descendants, and all generations of the future. The concept of intergenerational equity or *environmental justice,* from the human point of view, asserts that we owe something to every other person sharing the planet with us, both those present and those yet unborn. But what exactly do we have to give?

The only things of value we have to give are our love, our trust, our respect, and the benefit of our experience. These are the essence of human values embodied in every option we pass forward—or withhold in every option we foreclose. When everything is said and done, all we have to give the children of today, tomorrow, and beyond are choices and some things of value from which to choose. The quality of those options is governed by the mutual support of the three pillars of sustainability.

The Three Pillars of Sustainability

Sustainability, as a term, was first launched into the arena of common discourse in 1989 by a United Nations publication known as the "Brundtland Report."[1] At first, it was treated with suspicion by businesses and others who perceived it to counter their interests in promoting economic growth. The definitions seemed esoteric and hard to pin down for people of practical bent. But then, the report was meant merely as a framework within which people could begin a discourse on sustainability and its implications for the world's future.

The three pillars of sustainability (ecological integrity, social equity, and economic stability, often referred to as "the triple bottom-line") are the most easily understood and thus commonly used frame of reference with respect to the subject of sustainability in all its varied forms. In its simplest form, any act or project must simultaneously meet appropriate criteria in the social, ecological, and economic spheres in order to qualify as sustainable, or as sustainable development. The triple bottom-line is often depicted visually as a set of overlapping circles.

Although debate can arise over the relative importance of each pillar in relation to the others, it is irrelevant, because their significance is mutually complementary—not competitive. The interaction of the three pillars (their joint support, if you will) is the key to real sustainability, even though applying the framework can certainly lead us to rank projects, activities, or even attitudes, as "more or less" sustainable. Nevertheless, it is our contention that the system is pragmatic in use—to strive for perfection in an imperfect world. It can be employed to improve current projects or ways of doing things, as well as to assess new proposals.

Understanding the Triple Bottom-Line

We are all familiar with the old saying, "The proof of the pudding is in the eating." Although the concept may be intellectually elegant, it has little use unless it helps us think about and shape real-world issues. In other words, it must contribute directly to movement toward a just and sustainable world. How does it work in that regard? Let's look at some examples.

"The Natural Step for Business," a program first instituted in Sweden and spearheaded in the United States by Paul Hawken and others, has led many businesspeople to consider all their activities within the triple bottom-line framework. In doing so, they demonstrate that profits can be made while simultaneously protecting the environment and supporting healthy communities. Many businesses ascribing to the Natural-Step program are small businesses with owners already sympathetic to environmental protection. Their success offers positive, grassroots evidence that meeting material needs does not have to destroy our natural-resource base or our communities.

Many nonprofit, public-interest groups (such as the Worldwatch Institute, Earth Policy Institute, Positive Futures Network, and Union of Concerned Scientists) focus on the environment, resources, and energy and incorporate the triple bottom-line into their organizational philosophy. Included are organizations that spearheaded the whole approach in the first place, and they continue to be effective advocates.

Mainstream corporate business, however, is still responsible for the lion's share of the economic activity in our system, and practicality dictates that here must be the primary target if we are to have real impact. At first, there were perpetual calls from defenders of the corporate sector to define the term, which is a favorite ploy for innate skeptics and those prone to informed denial.

On one occasion, when I (RB) gave a talk on sustainable development to a group of people engaged in promoting economic activity, I received the comment, "Isn't sustainability just another thinly disguised way of getting us to put the environment ahead of the economy?" A reaction that surfaced on another occasion was that the reference to posterity and intergenerational equity in the common definitions of sustainability is an attempt to emotionally shame us into saving the environment by invoking our children and grandchildren, rather than promoting economic growth. I have even heard it referred to as, "A communist plot to destroy capitalism."

Whatever the specific nature of objections, they normally boil down to perceived attacks on the Growth Ethic, and then, by extension, on much that is accepted in our national story as patriotic and American. The emotional flag of the Land of Opportunity, with all the accompanying clichés, is readily waved.

On the one hand, questioning of the Growth Ethic is vital, because that pervasive element of our culture has proven to be leading us down a disastrous path. In our view, economic stagnation is virtually guaranteed under the Growth Ethic. Witness our current plight. On the other hand, social-environmental sustainability, comprehensively embraced and applied, can show the way to a healthier and more stable economy, a safe and vibrant environment with Earth-friendly technology, and supportive and just communities. That is to say, a sustainability-based view of the future is much more likely to preserve what the Growth Ethic purports to value and promote than is the actual pursuit of the Growth Ethic itself.

Over the last decade, the triple bottom-line has transitioned from a vaguely defined philosophical concept to one with more general credence. This has resulted from better articulation and increased understanding of its potential, as well as some successful, small-scale applications at grassroots levels, but nonetheless significant as demonstration projects. Increasing ecological and economic evidence that the current paradigm is leading us to social-environmental bankruptcy has bolstered all this positive reinforcement.

With this gradual movement into the mainstream current, another risk has presented itself. As always, corporate leadership is alert to anything that poses a threat to their procedures and profits. Certainly, a declining resource base and failing purchasing power would qualify as concrete threats. But before entertaining real change, informed denial will set in. At first, companies may fear *being seen* as unsympathetic to the idea of sustainability more than they fear *not being* sustainable.

The response to this risk in perception, and its cosmetic remedy, has been termed *greenwashing*, which amounts to active and vigorous public relations campaigns to convince customers and the general public that businesses ascribe to the principles of sustainability without actually making a substantive commitment in that direction.

We must admit that the three-pillars concept lends itself to easy philosophical agreement with little need for real commitment. In fact, a strong environmentalist mindset can articulately and vigorously insist on the implementation of these principles by calling for *others* to change their behavior. It can appear to be a way of saying, with no costs to oneself, "I am enlightened, and you should be, too." Businesses are quick to pick up on the fact that the burden of adjustment to a new paradigm falls largely on them. Risks of transition are certain to exist and threaten their way of operating, as well as their profits. Nevertheless, businesses—especially global corporations—are, and should be, on the front lines.

There are few easy answers to this blueprint of superficiality and social conflict. Our best response is that we need to assist moving the triple bottom-line from conceptual incubation into the arena of practical application. Rather than simply debate and admire the philosophical elegance of the three pillars (ecological integrity, social equity, and economic stability), we need to spell them out in operational detail and, in any way we can, to test what they mean and how they work in the practical, real world. We turn now to that task.

Ecological Integrity

Most people speak of stewardship when it comes to the health and biophysical integrity of our local, regional, national, and global environment. We think the concept of a *living trust* is preferable, because stewardship does not have a legally recognized beneficiary beyond the immediate owner—someone who directly benefits from the proceeds of one's decisions, actions, and the outcomes they produce.

Stewardship, therefore, is a much more restrictive term than *living trust*, because the fiduciary responsibility is only to the immediate "shareholders," whereas the fiduciary responsibility of a "living trust" is to all beneficiaries, many of whom need not be current physical shareholders—such as generations yet unborn. This more closely mirrors our concept of social-environmental-economic sustainability.

Biological Living Trust as a Management Tool

The concept of a biological living trust can effectively serve as a management tool for incorporating the functional aspects of ecological integrity into the triple bottom-line system. However, the notion still sounds somewhat esoteric. If, therefore, the concept of a biological living trust is to work, specific principles must be set out as guidelines:

- Everything, including humans and nonhumans, is an interactive and interdependent part of a systemic whole.
- Although parts within a living system differ in structure, their functions within the system are *complementary* (not competitive) and thus benefit the system as a whole.
- The ecological integrity and social-environmental sustainability of the system are the necessary measures of its economic health and stability.
- The integrity of biophysical processes has primacy over valuation of the economic conversion potential of a system's components.

- The integrity of the environment and its biophysical processes have primacy over human desires when such desires would destroy the system's integrity (read productivity) for future generations.
- Nature's inviolate, biophysical principles determine the necessary limitations of human endeavors.
- The disenfranchised as well as future generations have rights that must be accounted for in present decisions and actions.
- Nonmonetary relationships have value.

In figuring out how these principles would apply, as we attempt to assess a practical issue or project using the triple bottom-line, one point becomes quite clear: Ecology trumps economics, not the reverse. But, the fact is that the reverse has long held sway in our nation's culture and, consequently, led to the observed breakdown in both our economy and our environment. This is precisely what must change, and the triple bottom-line approach constitutes a step in that direction.

The Forest as a Triple Bottom-Line Example

Our system of national forests constitutes a biological living trust for all generations. A living trust represents a dynamic process, whether in the sense of a legal document or a living entity. Human beings inherited the original living trust—planet Earth—long before legal documents were invented. The Earth as a living organism is the ultimate biological living trust of which we are the trustees and for which we are all responsible. Our trusteeship, in turn, is colored, for better or worse, by the values our parents, peers, and teachers instilled in us, our experiences in life, and the ever-accruing knowledge of how the Earth functions as an ecosystem.

That said, the actual administration of our responsibility for the Earth as a living trust has, throughout history, been progressively delegated to professional trustees in the form of elected or appointed officials when and where the land is held in legal trust for the public—*public lands*. In so doing, we empower elected or appointed officials with our trust, our firm reliance, belief, or faith in the integrity, ability, and character of the person who is being empowered. With respect to a forest, for instance, we delegate to the U.S. Forest Service these responsibilities.

On public lands, such empowerment carries with it certain ethical mandates that are the seeds of the trust in all of its senses—legal, living, and personal:

- "We the people," present and future, are the beneficiaries; whereas the trustees are the elected or appointed officials and their hired workers.
- We entrusted these people to follow both the letter *and spirit* of the law in its highest sense.

- We entrusted the care of public lands (those owned by all of us), whether forest, grassland, ocean, or otherwise, to officials and professionals with a variety of expertise, all of whom are sworn to accept and uphold their responsibilities and to act as professional trustees in our behalf.
- Our public lands—and all that they contain, present and future—are "the asset" of the biological living trust.
- The American people have entrusted officials and professionals with our public lands as "present transfers" in the legal sense, meaning that we have the right to revoke or amend the trust (the empowerment) if the trustees do not fulfill their mandate: Public lands are to remain healthy and capable of benefiting all generations.
- Revoking or amending this empowerment if trustees do not fulfill their mandates is both our legal right and our moral obligation as hereditary trustees of the Earth, a trusteeship from which we cannot divorce ourselves.
- As U.S. citizens, we have additional responsibilities to critique the professional trusteeship (which includes economic oversight) of our public lands, because we are taxed to support the delegated trustees and also to provide public services from those lands. Elected officials make the dollar allocations on our behalf, and their decisions about where and how to spend "our" money are reflected in both the present and future condition of our public lands.

In other words, the process is necessarily ongoing as we speak. How might this work if we are to serve both as beneficiaries of the past and trustees for the future? To answer this question, we must first assume that the administering agency is functional, responsible, and governed by the concepts of social-environmental sustainability, of which economic integrity and social justice are integral parts. The ultimate mandate for the trustees, be they employees of an agency or otherwise, would then be to pass forward as many of the existing options (the capital of the trust) as possible.

These options would be forwarded to the next planning and implementation team (in which each individual is a beneficiary who becomes a trustee) to similarly protect and pass forward in turn to the next planning and implementation team (the beneficiaries that also become the trustees). In this manner, the maximum array of biologically and culturally sustainable options could be passed forward in perpetuity, all under the jurisdiction of consistent and informed public critique and the legal system.[2]

People with the necessary courage to unconditionally accept such a challenge and the change that it represents are rare, but I (CM) remember meeting one in 1992 in Slovakia. I had been asked to examine a forest in eastern Slovakia and give the people my counsel on how to restore its ecological integrity after years of abusive exploitation by the Communists. During the process, I worked with employees of the Slovakian Federal Forest Service. One man, the Chief Forester, then near the end of his career, had been in charge of the forest during the days of the Communists. As I was about to leave Slovakia, the Chief Forester took me aside and said, with great emotion: "Chris, if I learned one thing from you, it is that the forest is sacred—not the plan. Thank you." With that, this man reversed the thinking of his entire 40-year career. I have seldom encountered such courage, humility, and dignity.

We all need such courage, humility, and dignity if we are to be worthy trustees of our home planet as a biological living trust, because a living trust is like a promise—something made today but about tomorrow. In making a promise, we relinquish a bit of personal freedom with the bond of our word. In keeping that promise, we forfeit a little more freedom of action in that we limit our actual behavior. The reason people hesitate to make promises lies in the uncertainty of circumstances on the morrow. Helping to quell the fear of uncertainty is the purpose of a living trust.

Though a trustee may receive management expenses from the trust, the basic income from the trust, as well as the principal, must be used for the good of beneficiaries. In our example, the healthy and diverse forest is the capital, and under appropriate conditions, some trees can be considered interest on the capital to be used by current generations. In our capitalist system, however, natural resources are assumed to be income or revenue, rather than capital. Capitalism in practice has a way of drawing down capital and thus precluding options. This is a complete irrationality even by its own internal methodology, which holds that capital is to be preserved and enhanced, and not dissipated. That said, a true trustee is obligated to seek ways and means to maintain or enhance the capital of the trust—not to diminish it. Like an apple tree, one can enjoy the fruit thereof but should not destroy the tree.

For economics to survive throughout the 21st century as a constructive and useful profession, it must accept the moral essence of a biological living trust. It means we must think in terms of potential productivity instead of constant production. It must also advance beyond resisting change as a condition to be avoided (clinging to the current, linear, reductionist, mechanical worldview of competitive exploitation in an effort to fulfill the Growth Ethic) and embracing change as a dynamic process filled with exciting opportunities for the present *and* the future—the beneficiaries.

In a biological living trust, the behavior of a system depends on how individual parts interact as functional components of the whole—not on what each isolated part is doing. Thus, to understand a system, we need to understand how it fits into the even larger system of which it is a part. Consequently, we will move from the primacy of the parts to the primacy of the whole, from insistence on absolute knowledge as truth to relatively coherent interpretations of constantly changing knowledge, from attempting to solve old problems with old thinking to creating new concepts tailored specifically to the current, changing social-environmental context, and finally, from an isolated personal self to self in community. We now turn to that social, or community, element.

Social Equity

The pillar of community, or social justice, is the most difficult to define and thus the most easily ignored of the triple bottom-line. Economics and environment are more easily grasped, and analysis about them can be externalized. "They" must do something about that. "Technology" must find a way to deal with an economic situation, an environmental problem, or even the connection between the two. Nurturing a sense of community or assessing issues involving justice and fairness, however, demands consideration of personal values and introspection, which is much less comfortable for most economically oriented people.

We are, today, far removed from social justice. A wonderful example of perceived inequality among humans took place in a small Oregon town a number of years ago. It involved people living in the city versus those living in the country. An article in the local newspaper said a farmer had been arrested and fined for throwing garbage on somebody's lawn in town. As I (CM) remember reading the article, the story went something like this:

> Joe City, who lived in town, took his garbage out to the country and dumped it on Bill Rural's property near Bill's house. Although Bill did not see Joe dump his garbage, he found an invoice in the garbage with Joe's name and address on it. So Bill picked up all of Joe's garbage and drove into town, where he dumped it onto Joe's front lawn. Joe went to the police and complained.
>
> Even though Bill said the townspeople were continually dumping their unwanted garbage on his land and that in this case he was sure it was Joe's garbage, Joe had legal standing and Bill did not. Bill was arrested and fined, but *nothing* happened to Joe.

This sent a clear message of inequality to Joe, to Bill, and to everyone else. The message was that it is okay for city folks to dump their garbage with legal impunity on the property of rural folks, but not vice versa. It was apparent that, despite the Constitution of the United States, some people are a lot more equal than others.

Looked at another way, this example allows us to uncover some profound and universally applicable conclusions. Rural people who value clean air and good-quality water have a right to enjoy these amenities, especially when they purposefully live out "in the middle of nowhere." But bureaucrats, hundreds of miles away, give cities and industries the right to pollute air and water because of economic and political power. They do this despite the fact that such pollution fouls the air and contaminates the water that rural people use.

It could be said that because Joe lived in a city, his community supported him and his personal rights more definitively than they supported

Bill's rights, because he operated as an individual out in the country with little or no defined community and thus without the political power that accompanies such a social organization. Clearly, such a case is testimony to the fact that a social organization can serve as easily to promote inequality among people and protect privilege (and thus *injustice*), as the reverse.

It is not an unreasonable stretch to note that cities throughout human history have nurtured the kinds of social organizations that have fomented wars, conflict, and political and economic domination of the most egregious and unjust empires the world has known.

Although cities are often memorialized as the breeding grounds for the finest in human achievements—poetry, music, the arts, great architecture, and profound philosophical thought—they are also the fountainhead of tyranny, oppression, slavery, and inequality of all kinds. There are no guarantees.

Today, the simple fact that food is grown in the country and consumed in the city offers (as does the story of Joe and Bill) a telling metaphor: Real support for our existence lies in one area and the actual power and control in another. Capitalism and the market economy would seek to disguise this reality. City people supposedly support farmers by purchasing their products—rather than the other way around.

Cities and their corporate "henchpeople" finance rural community activities and socioeconomic health. Cities are touted as providing jobs that are vital for economic health. It is commonly assumed that environmental protection in general rests on the financial ability to achieve it, as if it is something that needs to be purchased. This is all true only within the alienating and self-centered construct of today's industrial nations and their consumerist societies. All good things supposedly spring from a market economy committed to the growth ethic.

The toxicity of this image is underscored by the recent financial and economic meltdown that was perpetrated in and around New York, our biggest city (a national and world financial center), which has negatively affected the real well-being of virtually everyone throughout the country. It becomes more apparent daily, as Wall Street bounces back (while we on Main Street do not) and continues to accumulate more wealth and power, that the real upshot of the entire experience is increased inequality and social injustice—the antithesis of what is required for a sustainable world. Perhaps capitalism is also the direct opposite of what is required.

These examples and observations, more than anything else, are a comment on human nature. The gene pool is rich, and humans are capable of all kinds of instinctual behavior—positive and negative. People are, however, deeply affected by their institutions; and the challenge, as we see it, is to nurture the development of political and economic structures that are more conducive to sharing, social justice, and a healthy sense of community than those currently dominating our lives. But the relationships are systemic and work both ways in that people both created and can affect the institutions as well. The process, therefore, must begin with personal awareness.

Human inequality has to do with fear and its economic companion, control. The person who harbors the most fear also harbors the greatest need to be in control of his or her external environment; the need to be in control is always fed by the need for the "inequality of enemies" from which one can (presumably legitimately) steal individual rights.

For example, I (CM) spoke at a forestry conference in Victoria, Canada, in March 1988. While there, I listened to an eloquent speech by a hereditary chief from a community of aboriginal Canadians on his people's right to the land on which they lived because they had never signed a peace treaty. Their traditional land had simply been wrested from them. To my complete astonishment, an indignant timber company executive wanted to know what right *they*—as Indians—had to own any land or cut any timber.

Inequality, which is simply another word for injustice, carries over into every institution in our land; but it is perhaps clearest in those agencies whose missions are to uphold and fulfill the legal mandates of protecting environmental quality for all citizens, present and future. Often, decisions about environmental control—or lack of it—bend to the political pressure of the economic elite at the expense of society as a whole, present and future. Decisions about the commons (such as clean air, pure water, and fertile soil, which are everyone's birthright) can create more injustice than in any other arena, because intergenerational equity is immediately and innately involved. There have been times, however, when equality and justice counted for something; as Thucydides said of the Athenian code, "Praise is due to all who . . . respect justice more than their position compels them to do" (Book 1, Chapter 3, page 75).[3]

It is now the beginning of the 21st century, a century in which once-abundant, natural resources are rapidly dwindling toward scarcity while the world's human population grows at an exponential rate. We citizens of this planet must now address a moral question: Do those living today owe anything to the future? If your answer is "No," then we will simply repeat the 20th century to the everlasting, progressive, living damnation of each succeeding generation.

But if your answer is "Yes," then we must now determine what and how much we owe future generations, lest our present collision course continue unabated into the future, eventually to destroy environmental options for all generations to come. But what direction must our renewed sense of personal and social justice take? We are not without historic guidelines.

Each great civilization has been marked by its birth, maturation, and demise; the latter brought about by uncontrolled population growth that outstripped the source of available energy, be it loss of topsoil, deforestation, or economic ruination due to avarice. But, in earlier times, the survivors could move on to less-populated, more-fertile areas as their civilizations collapsed. Settlement and population of the North American continent, along with the frequently cited national story of the founding of the United States, are our most common references to the firm, cultural belief in the omnipresence of the "next frontier" and "manifest destiny." We in these United States are the best example. Today, however, there is nowhere left on Earth to go.

Yet, having learned little or nothing from history, our civilization is currently destroying the environment from which it sprang and on which it relies for continuance. Civilization as we know it cannot, therefore, be the final evolutionary stage for human existence. But what lies beyond our current notion of civilization? What is the next frontier for "civilized" people to conquer? Is it outer space, as is so often stated? No reasonable person believes that is possible, especially given the short time we have with any modicum of available resources. What then? It is inner space, the conquest of oneself, which many assert is life's most difficult task. As the Buddha said, "Though he should conquer a thousand men in the battlefield a thousand times, yet he, indeed, who would conquer himself is the noblest victor."[4]

In the material world, self-conquest means bringing one's thoughts and behaviors in line with the immutable biophysical laws to which we have often referred. In the spiritual realm, this means disciplining one's thoughts and behaviors in accord with the highest spiritual/social truths handed down throughout the ages. It can be stated as simply as love your neighbor as yourself, and treat others as you want them to treat you.

The outcome of self-conquest is social-environmental-economic sustainability, which must be the next cultural stage toward which we struggle. Social-environmental-economic sustainability is the frontier beyond self-centeredness and its stepchild, *destructive* conflict, which destroys human dignity, degrades an ecosystem's productive capacity, and thus forecloses options for all generations.

The necessary adjustments will not come easily or smoothly, because "A great many people," as American psychologist William James observed, "think they are thinking when they are merely rearranging their prejudices."[5] Change, on the other hand, is about choices, those with which we either free or imprison ourselves.

Finally, as a practical matter, and derived from conclusions reached throughout this work, we offer some characteristics to look for and promote in the real world which would empower you in helping to create a community and an economy where equity and justice can thrive:

- Produce for your own necessities for local use—not just for export.
- Consider small-scale operations as vital as large scale operations.
- Promote local ownership and control.
- Use locally available resources, including workers, before searching elsewhere.
- Value those tools that enhance a worker's power over the machines, which replace human usefulness and ingenuity.

- Create institutions and business opportunities that allow communities to finance their own efforts.
- Emphasize economic institutions and arrangements that stress and highlight economic interdependence.
- Always address questions of distribution and justice first, regardless of the economic or environmental implications.

This is just a representative sampling of what could be listed. They appear to be economic questions, but they lead immediately to the issue of distributional equity and social justice, which will provide the glue for healthy personal and professional relationships in a vibrant and mutually supportive community.

We all live in a community somewhere. Practical issues and potential projects arise daily. It is our considered belief that features such as those noted above would assist in the task of creating an economics of stability, a large part of which is social justice. They will also help you connect with the following coverage of the third pillar of sustainability.

Economic Stability

Economics is the core of the third pillar of sustainability, the first two being *environmental integrity* and *social equity*. Although we have made the point that all three are equally important, it is appropriate to discuss economics last. Most physical, real-world projects that offer examples against which we might test the social equity criterion are economic initiatives, because they employ resources in an effort to support human beings.

In the world of the growth ethic, however, economics comes first. The doctrine of *economic feasibility* (narrowly conceived from the sole viewpoint of private profit) is an acid test. If the profitability test is passed, the product or process will proceed, even though it may have negative effects on the environment or on communities. If, on the other hand, it is deemed as narrowly unprofitable, even exceptionally positive implications for either the biosphere or social equity in general will not be enough to save it from the discard heap.

It is, therefore, appropriate to consider examples of economic activity, and thus test the effectiveness of the triple bottom-line framework. It is our intention that such examples will clearly demonstrate the mutual foundation of the three pillars. After all, a reasonable mindset is that economics provides the practical applications, and the environment and the community establish the constraints. Together, the three address the following question: How can humans support themselves in a manner that not only protects and maintains the resource base but also nurtures healthy and positive relationships for the quality of life that we all desire and deserve? This is the essence of socioeconomic sustainability, and it summarizes the gift we, in the current generation, owe to all future generations.

The first task, however, is to identify the necessary characteristics of the economy. We can think of no more appropriate place to start than to summarize the main points in our broad critique of current economics, which made up Section II of this book. As that critique proceeded, the phrase *economic stability* was never mentioned. If this broader agenda of stability is now invoked, the results of the critique might take on more meaning, in that economics, so long left unchecked, is forced to serve positive human purposes. It is time to force this powerful body of thought to yield the world we want. As someone wisely said, *Economics makes an excellent servant, a poor master, and a miserable religion.*

Consequently, a return trip to the major conclusions of Chapters 3 through 8 promises to pay excellent dividends in providing practical guidelines for any economic assessment. Elaboration is not necessary, because it has already been ample.

- Economics must be recognized as a *social science*, which incorporates a full range of human emotions and does not treat people as maximizing machines.
- Consumption must be viewed as necessary for comfortable survival, and not as an end in itself. Consumption beyond that should properly qualify as *wasteful desire.*
- Production must first adequately meet livelihood requirements and secondarily satisfy unnecessary wants. Profit should not simply follow an ability to pay.
- External, unintended effects of economic activity must be fully accounted for, incorporating the best scientific information available. Excessive uncertainty or effects on innocent parties must be grounds for stopping projects or activities.
- Equity and fairness must dominate efficiency in the name of pure profitability. Consideration of distributional effects must rank higher than strict financial feasibility.
- The macroeconomic picture clearly reveals a resource-stressed planet. The world's abundant human labor must be used to its maximum, and capital, which represents the environment and its limited resources, must be used sparingly.
- Overall, the growth ethic is a cultural phenomenon that is impossible to sustain and must be removed from mainstream economic thought.

These guidelines represent only a partial list, albeit an important core, of the changes necessary if *economic stability*, the third pillar, is to serve as a compatible working partner with the other two pillars, *ecological integrity* and *social equity*. Bringing about these changes will be a tall order and implies nothing less than a revolution in economic thought. Obviously, an integral part of such a revolution is political, given the necessarily serious adjustments in the decision-making processes. Corporate capitalism, as we know it today, can hardly be expected to be compatible with the results.

Imagine an economic evolution toward a system that operates according to rules similar to the principles stated above and at the end of the previous section on social equity. The guidelines and constraints offered by ecological integrity and social equity will be of critical value, and it would not be unreasonable to presume that a newspaper report on the process of adjustment could read: *Ecological integrity and social equity harness economics in the interests of comfortable viability for the human species.*

This sounds like a grandiose statement, but in truth we should settle for nothing less. Even so, getting there will be a long journey, and the scenario cries out for practical examples. Let us proceed to examine a few real-world proposals for the purpose of experimenting with how the use of the triple bottom-line concept might apply in practice.

Proposal to Build a Coal-Fired Electrical Generator

We clearly have need of electrical energy in the future, and coal-fired plants are thought to be among the least expensive methods of producing it. Although economics, acting alone, would rank such a proposal quite highly, environmental considerations elevate the cost considerably due to extreme externalities, such as pollution of the air with sulfur dioxide and nitrogen oxides that kill forests, which not only affects all life dependent on the forests but also the ability of forests to sequester carbon and thus act as a control for climate change. Moreover, coal, which appears to be abundant, is a depletable resource and thus fails the test of stability, even without the negative environmental effects. No one would want a coal-fired plant in his or her community; thus, it would qualify as an alienating technology that destroys, as opposed to supports, a sense of community with good-quality air. Substantial capital is needed to build such a plant, so only a large corporation will undertake building one. Without attempting to indict capitalism, we contend that this project will also negatively affect social equity through the uneven distribution of income.

Therefore, although the pure economics, discounting the externalities, would look favorably on such a project, it would soundly fail the test imposed by the pillars of ecological integrity and social equity. Incidentally, any large, centralized generating facility would probably be evaluated similarly, regardless of the specific energy source—with the possible exception of solar energy. Let us continue in the energy sphere with another example.

Establishment of a Solar Photovoltaic Manufacturing Plant

Here, the final product is also in the form of electricity, but with a difference. The electricity from a coal-fired plant is consumed impersonally from the grid, while solar panels can be individually purchased and put on one's home. There is more of an apparent relationship between producing and consuming. Environmentally, the ultimate source will not run out and does not degrade, so the proposal meets standards for economic stability. Socially, any community would take pride in the jobs provided by such a manufacturing plant, whether the product is installed locally or exported. Local installation, to the degree that happens, would not only create jobs and a sense of interdependency among local producers and consumers but also self-sufficiency in meeting our own needs as a community or region.

Economically, photovoltaic electricity is presently more expensive than that generated by coal, and a hard-headed capitalist would be quick to point that out. This leads to a fundamentally important point. The process of promoting socioeconomic sustainability must allow for the positive benefits of the social and environmental spheres to comfortably offset the economic drawbacks. The three must work as a system, which is exactly how the triple bottom-line must work. We leave this example with a hypothetical question: What would the decision-making process in your community have to look like if such a proposal were to be readily accepted as the best alternative?

Locating a "Big Box" Department Store in a Community

Communities everywhere have gone through similar debates about the pros and cons of locating a retail outlet of a major national or multinational chain in their town (can we say "Walmart?") Local issues concerning land-use planning (traffic, etc.) always come up, and occasionally even the broader economic implications of such planning. What can we learn from the triple bottom-line concept about such a situation, and in turn, what does such a situation teach us about the practical use of these tests for assessing sustainability?

Such stores are clearly profitable and therefore almost automatically receive the stamp of approval of mainstream economic thinking. The consumerist paradigm views superabundant availability of all goods as a positive thing, as well as the jobs in the retail outlet—despite their low pay and poor working conditions. Deeper reflection reveals some problems, however. First, the typical range of products caters more to desires than to basic necessities. Second, pride is taken by such companies in offering whatever is sold at the lowest possible price, with little regard for quality. Third, these corporations are noted for paying their workforce as little as possible so that prices can be held to a minimum. People make sacrifices in their producer lives in order to emphasize their consumer lives. This is not balanced, and it ignores economic interdependency. Finally, production of the goods in question occurs on a mass-industrial scale at some point—normally outsourced to a foreign nation—far away from the point of actual final use.

Turning to the pillar of ecological integrity, the impacts of the production methods are difficult to ascertain, although major multinational corporations having goods produced for export are frequently seen as choosing production venues for having the *most lax* environmental standards, in addition to the widely acknowledged global search for the lowest wage rates. Thus, the methods of production are likely to be under relatively harsh conditions, both for workers and for the local as well as the global environment.

Another important environmental factor stems from how energy is used when goods are mass produced, because they must be shipped worldwide to distribution facilities, normally by ship, thereby polluting the oceans of the world. Subsequent transportation from the distribution facilities via truck is the norm. The model of the big-box store encourages the use of fossil fuel, the automobile culture, and suburban sprawl. In the end, however, the consumer, by definition, is responsible for his or her own home delivery from the store. The company keeps the private benefits, or proceeds, and many of the costs are "socialized" out to the environment, the workforce, local governments, and the consumers. In summary, this is a classic example of externalities in the mode of Chapter 6, and it graphically displays the kinds of distortions that can result.

Consideration of the community effects is equally enlightening. First, anything that encourages the mass-consumption, auto-oriented culture is alienating to individuals and families. It encourages them to ignore the connection between their functions as producer and consumer, and the awareness of who supports them materially and who they support economically. They are encouraged merely to be price seekers who ignore quality and offer no concern for the workers who either produced the goods or are employed by the retail outlet.

In summary, joint consideration of such a proposal from the worldview of the three pillars of sustainability casts a remarkably negative shadow over such a proposal. Even the economic aspect, which can be termed its strength, is seen to have serious drawbacks to big-box stores when the wage/price implications are comprehensively considered. Finally, let us examine a different type of proposal, a community garden.

Establish a System of Community Gardens

Food purchased at Safeway is assumed by many in this culture to be cheaper and more convenient than going through all the hassle of growing their own. It is an enlightening exercise to use the triple bottom-line framework in asking the following question: What is the difference between what is going on in the mind of someone who agrees with that, and someone who disagrees? To approach this question, we consider a proposal to allow a community garden on a piece of vacant land owned by the town or city in which you live.

Economically, the cost must be assessed in various ways. The direct dollar flows are no doubt less than purchasing the same amount of food at the store, but if the gardener is careful to keep track of time spent sowing, weeding, watering, harvesting, and preparing the food for the table, the "cost" rises astronomically. In fact, suppose the economically astute gardener keeps such data and then computes an average wage rate for their efforts. In all likelihood, either the implied wage rate will be miniscule, or the implied dollar cost will be considerably higher than the grocery bill at Safeway would have been. So the market test is problematic. But, on the other hand, what if the gardener simply does not have the money in the first place or wants locally grown rather than that imported from Mexico?

Environmentally, several perspectives emerge. The Safeway food was normally grown via an agribusiness operation in a distant place, often in another country. Thus, chemicals and pesticides were probably applied, and considerable fossil fuels were used on the farm and in distribution and transportation to get them to market. The transportation costs of growing your own food amount to the pleasant walks to and from the site as you connect with the pillar of ecological integrity while you shepherd the wondrous transformation of seeds to ready-to-eat produce.

On the other hand, the average molecule of food in our current grocery system has been estimated to travel over 1,900 miles before we eat it, and judging by the taste, it was a long, hard trip. Therefore, the food is produced to lengthen its shelf life, in part by genetic engineering and in part by harvesting it green, at the expense of its flavor and health benefits. But the food you grow is higher in quality and nutritional value, while being significantly lower in the use of energy, both in the production and use of petrochemical compounds and long-distance transportation. The act of undertaking the entire process of a garden—especially organic gardening—teaches patience and the critical importance of soil health and the availability of good-quality water, and it makes one aware of climate and seasonal rhythms. We have no hard evidence, but it is our expectation that devoted gardeners—especially organic gardeners—are more aware than the average citizen of the possible effects of global climate change. People learn many things from their gardens.[6]

Socially, community gardens, along with their "companion features" of Saturday markets, are yielding amazing results in communities everywhere. The community garden becomes a meeting place—a commons, where people can share and discuss common interests and learn new skills from one another. ("How do you grow such good carrots? Let me show you my tomatoes.") Children can learn where food comes from and how much work it takes, and everyone appreciates a sense of interdependency, even as they promote regional and community self-sufficiency. Few gardeners do not take real pride in sharing the bounty at harvest time—the very essence of community. Further, any efforts to preserve food for the winter or other out-of-season use will enhance one's sense of long-term planning and the difference between sufficiency and surplus.

The Safeway customer, on the other hand, has no such opportunities for learning, awareness, and personal growth, but rather seeks to grab a bite to eat and get on with his or her specialized job in the money economy, perceiving that the customer does not have time for such "rural" activities. Although such a person will not readily perceive this, compared to community gardeners, this whole approach is actually quite isolating, even alienating, not only from one another but also from the feeling of true community, which is based on mutual caring.

As a final note, the reader would be justified in complaining that these examples seem to be "straw men" or "Trojan horses" set up not to represent actual choices, but to allow us to make a predetermined case for certain activities or for the use of the three pillars of sustainability. Our first response is "guilty as charged." We do wish to use clear-cut examples to demonstrate how anyone can use the framework in testing the effectiveness of a proposed allocation of resources. And perhaps the best examples at the outset are clear-cut situations.

However, we set out to make an important point. Anything a person, a business, a community, or a nation is doing now or sets out to do can be assessed effectively using this system. Although it is admittedly the case that the resulting evidence, and therefore the judgment rendered, will not always be as clear-cut as in these examples, the results of the analytical effort are guaranteed to enlighten. Our purpose here is not to recommend that you, the reader, buy solar collectors and garden (though we think you would enjoy them), but rather that you experiment with using the triple bottom-line framework, individually and collectively, in assessing projects and activities that affect all of our lives. In fact, the use of the process makes an excellent community-building activity. We cannot imagine a more constructive event than a community workshop employing the three pillars (ecological integrity, social equity, and economic stability) to assess the proposal of a real project of interest to the entire community.

With practice comes clarity about where we are headed as a culture and about what we may need to do to insure a fulfilling life for our loved ones and ourselves, and to insure that future generations enjoy the same opportunities with which we have been so abundantly blessed. It is all about living in long-term harmony with our planetary home. We think the triple bottom-line framework can assist the achievement of such a goal, and that is a powerful contribution to social-environmental sustainability for all generations.

Sustainability in Practice—The Track Record

The idea of the triple bottom-line has been around for some time now. Many individuals, groups, and organizations have—to their credit—attempted to employ the concepts in actual practices and projects. We are compelled to ask: How is it going? What have been the results, the successes, the shortcomings?

What can we learn from these early pioneers that will improve the effectiveness of a process we hope will become pervasive and permanent?

Focus on the environment has been palpable for at least 40 years. Many of the environmentally related nonprofit organizations and public interest groups have found the triple bottom-line framework an appealing and compatible ally. Often, such groups have come into existence to oppose some particular type of economic activity by taking up a particular cause. Greenpeace and Save the Whales oppose factory-fishing fleets and the incidental catch of nontarget species. The Union of Concerned Scientists argues against the efficacy of nuclear power, and opposes offshore oil drilling, as does the Environmental Defense Fund.

Worldwatch Institute has long championed images of a sustainable world and regularly publishes separate pieces on forests, oceans, farmlands, freshwater resources, climate change, energy options, and a host of related topics. Whatever their particular point of view, the common denominator supporting the efforts of all these organizations is a critique, overtly or tacitly, of economic activities in light of ecological integrity and social equity. When alternatives are presented, they are universally more compatible with triple bottom-line criteria than the current economic activities they are opposing.

This is all very encouraging and represents real advancement in the arena of public debate, but there is one problem. The conversation typically focuses on the connection between economics and environment, and says little about communities and social justice. The technological imperative of this culture of growth is a powerful force, and it is very hard to alter, even in the minds of those who strongly perceive the need to change the current paradigm. Let us expand on this through another example.

The Green Building Council promotes use of sustainable materials and energy-saving techniques in the construction of housing and commercial/industrial facilities. The LEEDS program (Leadership in Energy Efficient Development) symbolizes this. To be sure, the results are touted as being more "livable" for families, but mainly because the monthly dollar costs (after perhaps a considerable capital investment) will be lower. You are told that you can have all the comforts you had before. In other words, you do not need to give up anything or adjust your behavior—technology will see to that. Personal and family sacrifice are not required, and community solutions are rarely mentioned.

As a result, many proposed solutions are reasonable only if you can afford them. And the devotion to technological solutions remains intact. As we will discuss in the two concluding chapters of this book, the need is for community-based bottom-up ideas and solutions that will truly speak to the disparities between income and wealth and promote healthy relationships among people. All many people really want or need is a sense of belonging and being valued in a true community, one built on mutual trust and caring. Accordingly, the most pressing need in the advance of triple bottom-line thinking is for more integration of the social-equity element. If you recall, the economics critique of Section II singled out distribution as the biggest problem our system faces. In our experience, the three-pillars methodology reinforces that contention. Integration of economics and environment has progressed rapidly, thanks in part to its innate reliance on technology—especially energy technology. Social equity (the "people part") must be emphasized as we move forward.

At the very least, this cursory assessment, offered without systematic or comprehensive data, could help to further familiarize you, the reader, as to how the process could work. As always, examples are instructive. Look around you. Everyday activities in any community provide ample opportunities to make observations and comments such as the following: "Oh, I see the strengths and weaknesses of that proposed project now. Here's how we could improve it. Or, how about this as an even better alternative? And here's why."

As common citizens, we are conditioned by the consumerist paradigm to feel powerless. (What influence could I have? I am just one person.) We feel so alien from those we support and from our own sources of real support, that we may as well just seek personal fulfillment through goods and acquisition. It is a failed paradigm, and it has had nearly disastrous effects on our world, our communities, and our own thinking. Pay attention in your community. Talk to others. Think. Get involved. Your time will come. Your gifts of insight and talent are desperately needed—share them.

Endnotes

1. The Report of the Brundtland Commission. *Our Common Future*. Oxford University Press, Oxford, England. 1987.
2. The foregoing discussion is based on: Chris Maser. *Our Forest Legacy: Today's Decisions, Tomorrow's Consequences*. Maisonneuve Press, Washington, DC. 2005.
3. *Thucydides, the Complete Writings of*, the Crawley translation, introduction by John H. Finley, Jr. The Modern Library. Random House, New York. 1951.
4. Buddha. *The Teaching of Buddha*. Kosaido Printing, Tokyo, Japan. 1985.
5. William James. www.brainyquote.com/quotes/quotes/w/williamjam109175.html (accessed on December 10, 2010).
6. Chris Maser, with Zane Maser. *The World Is in My Garden: A Journey of Consciousness*. White Cloud Press, Ashland, OR. 2005.

10

Imagining the Ideal World

In the previous chapter, we set out to develop the criteria for assessing whether an individual is on the right track with his or her thinking, actions, and consideration of projects and activities for his or her community, assuming the individual desires to help build a sustainable world. In this chapter, we seek to develop a possible vision of that sustainable world, one built on the foundational principles encompassed in economic stability within the triple bottom-line framework.

As important as it is to have guidelines against which anything can be tested, it is equally important to have a clear idea of where you are headed. What does success look like? How will I know it when I see it? Are there positive models that can be constructed or, better yet, examples that already exist? Addressing these types of questions is the objective of this next-to-the-last chapter. Finally, in the last chapter, we conclude by offering whatever counsel we can for taking practical action.

In a real sense, the criteria for a vision of sustainability based on the three pillars will continue to emerge in cumulative fashion throughout the rest of the book. Much of what might qualify as vision, however, has already appeared. Overall, it is our purpose to coalesce how one might go about achieving a world of social-environmental-economic sustainability by using positive, concrete images to nudge your imagination of what is possible.

We perceive a delicious irony in undertaking this task. Throughout, our message has been systemic, wherein everything is connected to everything else and must be considered as a whole if any part is to be understood. Because much of the topic is ecological—the science of understanding interconnectedness—this is not surprising. As authors, our inclination is a desire to say everything at once, because our perception of the message is of a fully fleshed vision. But that is not the way books work; they are necessarily sequenced and linear. Therefore, we must talk about thing A in a certain place and thing B in its place, no matter how connected they are. Bear with us, dear reader, as we struggle to help create the new paradigm, while honoring the best and most workable of the old. But, first it must be understood that we cannot move away from a negative.

We Can Only Move toward a Positive

Every enterprise needs to have the organizing context of a vision toward which to strive, be it an entrepreneur building a business of ecotourism or a community planning for its future. This need is particularly true of a community, which must create and work within a shared framework—the triple bottom-line. As a strong organizing context, a shared vision has some distinctive traits:

- It tends to focus a wide range of human concerns.
- It is strongly centered in the community.
- It can use alternative scenarios to explore possible futures by depicting in words and images that which a community is striving to become.
- Its creation relies on the trust, respect, and inclusivity of interpersonal relationships.
- It is ideally suited to, and depends on, public involvement.
- It is ideally suited to the use of creative, graphic imagery.

Although a shared vision does not replace other kinds of planning, it is the organizational context within which all planning fits (ecological integrity, social equity, and economic stability), a *positive* context that is all too often forgotten. Thus, the greatest single agent of failure to achieve one's desires, whatever they are, is not understanding the importance of a vision, how to create one, or a commitment to its implementation. This observation is *especially true of communities* and their commitment to the triple bottom-line of sustainability.

There is great power in learning to reframe negatives into positives. In doing so, the participants in creating a shared vision not only understand their community from several vantage points but also understand that much of the confusion in communication comes from trying to move away from negatives. Trying to move away from a negative precludes people from saying what they really mean because they are focused on what they *do not want*. As long as people express what they do not want, it is virtually impossible to figure out *what they do want*.

Although our educational systems in the United States, beginning with parents and ending with universities, stress the positive, they usually teach in terms of the negative. What does this statement mean? What might it cause?

Suppose your neighbor lives along a busy street and has a little boy named Jimmy. Your neighbor is concerned about Jimmy because of the increasing automobile traffic in the neighborhood.

One day Jimmy's mother says to him: "Jimmy, don't go into the street." The directive words (those telling Jimmy what to do) are *don't go* (a confusing contradiction), and the last word Jimmy hears is *street*; he thus follows the direction of the two congruous words he hears, *go* and *street*, and gets hit by a car.

What Jimmy's mother really meant and needed to have said was "Jimmy, stay in the yard." Then the directive word (the one telling Jimmy what to do) would have been *stay* (singularly clear and concise), and the last word Jimmy would have heard would have been *yard*. He would still have followed the two congruous words he heard, *stay* and *yard*, but with very different results—he would be alive.

This example illustrates that having been raised trying to make positive statements out of negative ones, we spend most of our lives trying to move away from the negative—and we cannot. We can only *move toward* a positive,[1] which includes *economic stability*, the third pillar of the triple bottom-line.

Economic Development in the Current World

Most of us have some awareness of efforts aimed at economic development in our community. Normally, the local press, given the siren call of increased advertising revenues, will readily tout that perceived need and any efforts to fill it:

- Get the economy moving again!
- Create jobs!
- CEO of XYZ Corporation visits city to consider locating a branch plant—provided certain concessions are made—negotiations confidential!

We have all seen similar statements, and in these days of a visibly failing economy, the calls are even more strident. Anyone close to (or seeking to be near) a position of influence seems to signal the need to promote higher levels of economic activity. Attracting jobs to "our" town is as American as motherhood and apple pie.

As an economist with expertise in regional and urban economics, I (RB) have both observed and been directly involved in many aspects of this pervasive process, including inventories of potentially industrial lands, zoning for location of business, creation of economic-development districts, and the calculation of spending and job multipliers for measuring the direct and indirect economic impact of some new business or activity. Having taught for 33 years at a university located in a state capitol (Willamette University in Salem, Oregon), I have even participated in writing legislation designed to "improve" the economy, and testified before legislative committees considering such bills. And I have come to question the effectiveness of most of these efforts.

This is the old model, and while structure of community power, from the Chamber of Commerce down, can appear to be behind it, at times nobody seems to have a clear idea of specific methods to achieve this holy grail of a vibrant, full-employment economy. Even though the goal is local economic activity (all business is located somewhere), the deeply ingrained top-down control will not let them forget the orientation to the national and global economy. Conservatives promote lower taxes and less regulation, despite the reminder that it is precisely this lack of self-control first and legal control second that caused the recent financial meltdown and housing crisis in the first place. Liberals advocate more public investment and stimulus spending, and they are attacked as "socialists." At critical junctures, especially when specifics on how to "fix" the economy are called for, both sides can fall strangely silent.

It is instructive at this point to suspend all the idle chatter and systematically examine the typical process of economic development as it has historically taken place over the years in cities, states, and regions of the United States. Then we will contrast what is typically done to see exactly how sustainable economic development might unfold within the triple bottom-line of ecological integrity, social equity, and economic stability.

Conventional Development Principles

Clearly, when we examine what we will call the *conventional model*, strategic details differ among localities and parts of the country, but some strong similarities and common assumptions emerge. As a beginning, some of the commonalities in stated goals and assumptions can be presented:

- Bring in jobs by attracting a major corporation to locate in your state or community.
- Employ the local labor force.
- Concentrate on export-based products you can sell to other regions (or ideally worldwide foreign markets).
- Make substantial tracts of fully serviced, "shovel-ready" land available.
- Preapprove the appropriate industrial zoning and infrastructure for those lands.

These are the principles and assumptions that typically dominate local efforts involving economic development. To sweeten the pot, some typical incentives include the following:

- Promote breaks in property taxes (e.g., often 5 years) on the industrial lands.
- Consider free or inexpensive "below-market" deals on these lands.

- Consider direct, financial grants to prospective companies.
- Offer to develop training programs in local educational institutions to train the appropriate labor force.
- Undertake trade missions to all corners of the industrialized world to promote the above.

Sound familiar? This scenario is typical of the marching orders deemed necessary to *become competitive with the global economy.* That mantra should sound familiar as well.

And what are the problems with this picture? In the recent era of rapid globalization, the efficacy of this overall development strategy for states and communities has been increasingly questioned. Such criticisms have been mild, however, because no one wants to be accused of being *antigrowth, antibusiness,* or not *community spirited.* In addition, and despite a feeling of general malaise that persists, critics of the conventional model rarely see a clear alternative. After all, you cannot fight globalization, can you? It is irresistible, isn't it?

Rather than simply reiterating some of the stated reservations, which are normally offered with few viable alternatives, we choose to address the implications of the above points specifically, as they might apply within a philosophy of the triple bottom-line and its three pillars—ecological integrity, social equity, and economic stability. In this manner, we critique the conventional approach, in which we see little possibility for constructive results but ample potential for negative outcomes. However, something as deeply ingrained in conventional wisdom will remain pretty much invulnerable to outside criticism unless those criticisms are accompanied up front with a better alternative. And this gets us back to the specific agenda of this chapter: *Create a positive vision for worldwide economic stability.*

As promised, we need to reiterate important points made earlier. It is now time to use them. In the previous chapter, at the end of the section discussing the pillar of social equity, we listed some guidelines for creating an economy wherein equity and justice can thrive. These are the key ingredients upon which we must focus, because, as we will soon see, the element of distributional fairness and grassroots-community health is most often missing with the current model. Again, the principles, for what we now choose to call a *local approach,* are as follows:

- Produce for your own needs and not just for export.
- Consider small-scale operations as vital as large-scale operations.
- Promote local ownership and control.
- Look to employ locally available resources first.
- Value tools that enhance a worker's power, over machines that replace their usefulness and ingenuity.

- Create institutions and business opportunities that allow communities to finance their own efforts.
- Emphasize economic institutions and arrangements that highlight economic interdependence.
- Always address questions of distribution and justice first, regardless of the economic or environmental implications.

Assessing Conventional Wisdom

There were 10 points describing the conventional model enumerated earlier: five principles or assumptions and five incentives that states and cities frequently consider. In light of the principles of the local approach just reiterated, we now assess the 10 points in turn.

1. Attract a Major Corporation

First, there are perhaps 10,000 other communities also trying to get a branch plant of any company willing to expand, but the company is headed for China or Mexico. And the head office is not moving, unless it is to the Bahamas or Dubai for tax purposes. Anything a community does get will be the absolute lowest-paying jobs in the corporation, and your community, in effect, becomes an economic colony. The decision making remains elsewhere.

With the local approach, new startups are the goal, and there is no effort whatsoever to woo big corporations. This alternative has its own challenges, however, which we will discuss.

2. Employ the Local Labor Force

Certainly, if a community succeeds in getting a major company to open a facility, they employ some local workers, but doing solely what the *company* wants or needs them to do. Despite normal talk about "future expansion," a major company often employs the maximum number of workers it will ever have when it first locates in that community. These days, a locally situated corporation frequently evolves into a perpetual "cutback mode," which is the common operating model for many corporations worldwide. After all, virtually every major corporation, domestic or international, has been routinely shedding local workers for decades, if for no other reason than outsourcing. Avoiding labor costs is often the prime concern, which we are all aware is the driving force behind outsourcing.

With the local approach, startup companies employ a certain number of workers at first, and, if successful, unerringly expand employment over time. Further, a range of job types will be available to fulfill the company's requirements. The average salary per employee is likely to be higher, because the entire operation, including the chief executive officer (CEO), is located in the community. CEOs of locally-based companies will not earn as much as CEOs of some multinational corporations, but they will earn an excellent living if the companies succeed and are likely to contribute as leaders or minor philanthropists in the community in which they live and which supports them. Growth of the employment base and rising average salaries are much more likely over time than with the conventional approach to economic expansion. Moreover, the decision-making power stays in the community.

3. Produce Products for Export

Invariably, traditional economic development is strictly targeted at firms that produce goods for export—out of that locality if not out of the country. Conventional, export-based economic wisdom discounts the value of the service sector with the euphemism: "After all, you can't just 'take in each other's laundry.'" Fundamental economic analysis is clear: No city, state, or region is a closed system. There are always goods you do not produce yourself, which you need to import from other areas (automobiles might be a good example). Such purchases send your dollars outside the area; goods that you produce and send outside to others are needed to reverse the flow and bring dollars back in for recirculation.

Facilitating production of export-based goods, therefore, is the historic role for large corporations. Consequently, attracting them is seen as the salvation of the local economy. However, we do not need more corporations to come in and simply help suck local dollars out of the local community. Walmart does enough of that for all absentee-owned corporations combined. Of course, an export-based strategy puts a local labor force in direct competition with labor markets all over the world, because a good produced for export can, by definition, be produced virtually anywhere.

A local approach differs starkly from conventional economic goals and models. The sales target is strictly the local market. There is no goal of producing for export. If a business should thrive, and its productive output begins to exceed local demand, the market could be expanded in concentric circles around the point of production. But, the consumer base must be kept as close to home as possible, because that is the most likely way to retain a loyal market and insulate the operation from the predatory competition of a national company or chain.

It is reasonable to ask how this helps the local economy, compared to a company that produces for export and thus relies on the income of other areas to support a healthy level of sales. The answer is simple. Export-based production seeks to bring in dollars spent on purchasing goods that we do not produce locally. A local approach seeks to produce and market those goods so the dollars never leave in the first place. The impact on the local economy is the same as for an export-based product, and consumers still get what they need of the goods in question, plus all the accompanying benefits discussed here. There is no more important point about this strategy than this one, which includes many of the other benefits as well, such as good-quality products, ecological integrity of the community's environment, social equity for its citizens, and economic stability for its businesses.

4. Have Industrial Lands Available

Most communities desperately seek to have tracts of land (as large as possible) available for industrial location, as if just having space will attract a large corporation. In recent decades, however, industrial land became scarce and prohibitively expensive in the burgeoning Silicon Valley in the area of San Francisco Bay. Consequently, watchword among communities in the Pacific Northwest that were desperately seeking an expansion branch of the proverbial "good, clean, high-tech electronics" variety was that these companies would not consider a site of less than 100 acres.

There are not many qualifying parcels in most communities, because, as already stated, the parcels must be flat and encompass at least 100 acres, which inevitably means farmland. It is useful at this juncture in the evaluation of the 10 points of conventional thinking to pause and explain the hidden consequence of this choice in terms of the biophysical principles that govern the outcomes of our decisions. The pertinent ones are as follows:

- *Principle 1.* Everything in the universe is a relationship supporting relationships, thus precluding the existence of an independent variable and thus absolute freedom from unwanted consequences.
- *Principle 4.* All systems are defined by their function, not their pieces in isolation of one another, which includes every economic system.
- *Principle 5.* All relationships result in a transfer of energy, in this case from growing food in perpetuity for a community to satisfying the economic desires of the few for immediate monetary gains at the expense of growing food year after year.
- *Principle 6.* All relationships are self-reinforcing feedback loops; whether "positive" or "negative" depends on whether or not they satisfy a human desire. The question in this case is to satisfy whose desire—the wealthy to increase their ability to turn yet a larger profit, or the poor for readily available, less-expensive local food.
- *Principle 7.* All relationships have one or more trade-offs. The primary trade-off of committing farmland to use as industrial sites is less land on which to grow food for a burgeoning human population. This decision carries with it a subsequent trade-off of higher food prices for those who can least afford them. And there is yet an additional trade-off of polluting the groundwater from the production of industrial waste—pollution that ultimately contributes to poisoning the oceans, which reduces their ability to feed millions of people worldwide.

- *Principle 9.* All relationships are irreversible because all outcomes are novel, and we cannot go back in time to recapture a past condition. Once a parcel of land is committed to use as an industrial site, it cannot or will not be reclaimed as farmland, first because of the expense of removing the artificial structures, second because the soil is polluted, and third because no one will spend the money to do so—they just walk away from a venture that is no longer profitable, and leave the consequences to someone else.

In addition to the above principles, the siren song to such a locality is enhanced by the information that they would not use all that acreage at first, but would create even more jobs through future expansion. In reality, data show that firms never used more than about 15 acres, and the rest is held not for expansion, but for speculation. One thing that "glamour" firms do understand is that the announcement of their impending arrival will drive up industrial-land prices in the area, a phenomenon made even more lucrative if a low-priced "sweetheart" deal can be arranged in the first place.

None of these games need be played with the local approach. First, a startup company may not need much land. Second, the entrepreneurs are already locals and are presumably aware of the entire range of alternatives in the local real-estate market, and—if they are astute—their community's need for food over time. They are very likely to be able to find just the right location at a reasonable market price. If later expansion is appropriate, that will probably be accommodated readily, and the whole process will put much less strain on infrastructure services of local municipalities.

5. Zoning and Infrastructure

Along with the *amount* of land, the conventional approach can impose strident demands in the area of *quality* of land. If the average, large corporation is to be attracted, the industrial zoning must be exact and must protect the firm not only from incompatible commercial or residential uses but also from more intense and potentially polluting uses by heavy industry. Of course, if a heavy industry unexpectedly shows interest, the municipality will often scramble to change the zoning, because any old smokestack will due in a storm.

Further, the nature of needed services to the site can be quite exacting. Electronics firms, for example, often have extremely high requirements for water, both quantity and quality. A local area can essentially be asked to sell or degrade its best natural resources in exchange for some jobs. A full range of services (police, fire, water, utilities, etc.) is expensive to provide, especially in advance of a firm commitment on the part of a company to establish itself on the promised site when no property taxes are yet available to support the expenditures.

Needs of this type should be dramatically lower under the local approach. A startup firm is in no position to force unreasonable bargains or subsidies, and whatever services that might be needed can presumably be negotiated among existing members of the community. In fact, the required land may already be owned, because existing business interests may well be involved in any new venture. Moreover, locally based owners and managers will be much more sensitive to potential environmental damage because they live there, raise their families there, and may already have a lifetime commitment to the community. In addition, they are vulnerable to the ire of their fellow citizens, with whom they already identify if they in any way destroy the community's social-environmental quality.

6. Property Tax Breaks

Breaks in the amount and timing of paying property taxes are one of the most common incentives to lure business to a specific location. Because there are literally thousands of communities willing to offer such incentives, for many large firms it qualifies simply as a necessary condition. Often, as with *enterprise zones,* these tax breaks are for a 5-year period. Common among corporations seeking to establish a branch operation (somewhere) is the premise that unless they can foresee a complete return on their investment in 5 years or less, it will not be deemed feasible. Then, as today's corporate thinking goes, having recouped their initial investment, they will need to make a decision whether to stay or leave. Especially with high-tech firms, whose technology changes so rapidly, it often proves feasible to simply abandon the old facility rather than upgrade it, especially with another community somewhere else waiting to start the next 5-year tax-free clock running all over again. (See principle 9 under point 4 above.)

These types of incentives will be much less prevalent under a local strategy. All their options, by definition, are properties in the local area. The apparent bargaining power is considerably less, and tax variations may not even be considered an important variable with the local interests involved. At least, if a tax break is offered, the increase in wealth that results will accrue to someone in the community, as opposed to having the benefits disappear somehow to the head office of some national or multinational corporation. It will definitely serve in some way to benefit the local economy. The subsidy stays in town.

7. Free or Inexpensive Land

Here, the dynamics are similar to those for property-tax breaks. Any such "sweetheart deal" goes to the bottom line of a balance sheet and is probably not kept in the local area. The local benefits sought in exchange for the tax break must be strictly in terms of the jobs created, because the economic impact of the subsidy disappears into the books of the corporation. Further, if any land appreciation or capital gains result in later years, those gains will not benefit anyone in the local area.

Similarly, a local entrepreneurial team is less likely to seek such a subsidy, although their need for it and potential productive use of it are probably greater. All gains in local wealth, including taxable incomes, remain in the community and its immediate surroundings.

8. Direct Financial Grants

In recent years, it has become a common practice to make direct, financial grants (often administered through the state in question) to firms that agree to locate in the area. Typically, only large corporations can qualify, because only they are assumed to employ enough workers to really make a difference. This is little more than legalized bribery, even though it may be in lieu of offering some public service or infrastructure. Rest assured that if some service is forsaken, it is either not needed by that firm, or the firm can provide it inexpensively. Again, this goes to the bottom line of the overall corporation.

Any direct, financial grants under a local strategy end up enhancing incomes, profits, and thus wealth in the local area, and will, in all likelihood, be handled with much more transparency. For instance, the funds could go to salaries, training programs, or purchases of capital equipment, all of which have long-term benefits for the area, may represent expenditures subject to local spending multipliers, and may even increase the tax base and thus represent a measurable return on investment for the granting entity.

9. Education and Training Programs

Cooperative partnerships between economic-development agencies and local community colleges are frequently forged for the purpose of assuring any potentially interested business firm that it can be readily supplied with a sufficiently skilled labor force or that one can be quickly developed. Although this type of investment in human capital is a good idea under any strategy, the problem with this factor and big corporate clients is twofold. First, they want to be at full production as quickly as possible, and any needed training program is probably not up and operating.

Second, as mentioned earlier, most corporate expansions seek to offer jobs at the very bottom end of their salary range and with little needed technical skills. You do not need to attend a local college to work in a communications industry call center, as many English-speaking people in India and the Philippines already realize. In addition, if a newly locating firm needs key, technically trained people, they are often brought into the area from other branches in the corporation. Any time a job is created that also requires a new person to move into the area, the net job creation benefits of the new business are lessened.

With the local approach, the opportunities for effective educational and training programs would be greater. The firms may start small, and a predictable expansion in the labor force can be smoothly served over time. The startup entrepreneurs may even have already attended the local educational institutions, and the close cooperation needed to develop effective training programs will be much easier, given that the business firm employs local people at the start. Additionally, the range of skills that can be usefully provided through local institutions will be broader and will include business management skills as well as any needed technical skills. The successful manager of a startup company may even end up teaching a course at the local community college, or acting as a consultant to another community seeking to replicate the success. Who knows?

10. Trade Missions

Ah, yes, trade missions. Some of them occur to open up markets and thus sell more of the products a state or region already produces. Others are to talk to corporations that are potentially willing to locate in the area. Conventionally, it is thought that the more far-flung the trip, the better. An envoy to California might not attract much attention, but a trip to China or the European Union should *really* jump-start the economy.

There is something a bit pathetic about the whole premise. It says, "We can't stimulate our own economy. We don't have the money or the skills. *Please, come save us.*" In effect, it is an invitation to offer all the features just discussed in the first nine points, whereby you are asking them to "pick your pocket." If the goal is to attract new firms, rest assured those companies are playing one community against another, and all the resources given up to land them represent a particularly cynical form of a zero-sum game.

To be sure, regions have occasionally succeeded in generating more sales of certain products, where the area is particularly good at or suited for something. Even then, in this era of strident globalization, some place in the world is often quick to ramp up production and seek to undersell whatever products the trade mission addressed, primarily through much lower labor costs. And on the general issue of opening up markets (e.g., China's 1.3 billion people), it is telling that after all the efforts of the last two or three decades to liberalize trade, we now import more than 10 times the goods from China than we export to them. How is the agreement, sponsored by the World Trade Organization, working out for you?

Summarizing the Local Approach

Clearly, none of this is an issue with a local strategy. There is no need to induce a firm to locate because it is a startup, and its key managerial people, as well as its labor force, are already local residents committed to the economic health of their community. What is more, they are already invested in and supportive of local institutions, such as schools, service organizations, churches, and local government.

Finally, there is no need to undertake trade missions to sell the product, because local residents are the targeted market. In fact, the targeted market includes the workers, the owners, the investors, and the whole network of family, friends, and acquaintances of all of the above. Moreover, they all realize that their purchases keep the local economy healthy, to say nothing of supporting the people they know and love. Realizing all this, the people comprising the local market would not consider purchasing that good anywhere else—the firm has a *de facto* monopoly that will expand organically.

In conclusion, this leads to what may be the most important point. Due to the lack of need for the extensive monetary expenditures for transportation, distribution, storage, advertising, and marketing which are necessary for a national or transnational corporation, *the local entity has a cost-based, comparative advantage that cannot be competed away*. This is the dream of all commercial operations. It also is the best possible strategy for a healthy environment, because, among other factors, energy requirements per unit of output will be much lower. Ironically, a major question that must be answered is as follows: Can we produce efficiently and effectively on a *small-enough scale* to make it happen? This is one question you will never see in a textbook.

Targeting the Strategy

The previous critique of the ways in which economic development has taken place in communities all over the United States clearly implies some unspoken assumptions about the nature of products and processes. To effectively pursue and complete the aim of this chapter, we need to flesh out the vision with some more specifics.

First, a product is chosen, and a business is started to produce that product for local markets. The product or products must be basic necessities, as opposed to more esoteric luxury items or frills. The business must be owned locally, financed locally, employ local people, and ideally involve well-known technologies. The perceptive reader will note that the ideas raised in this short paragraph (e.g., product, finance, business organization, technology, and market identification, for starters) are just about the whole ball of wax. Changing what we do in all these areas would constitute, collectively, nothing less than an economic revolution, and thus deserves elaboration.

Choice of Products

Basic necessities must be chosen because the provision of something needed by everyone is basically recession proof. Economic fluctuations will affect the enterprise less than they would some luxury item, which can be quickly given up in a pinch. By way of encouragement, we note that much of this is already happening with that most basic of necessities—food. Community gardens, community-supported agriculture, Eat-Local movements, Saturday markets, and a host of related efforts are springing up everywhere. Included are restaurants committed to purchasing only locally grown products whenever possible.

Energy is another necessity whereby we are urged to do more locally. For example, implement conservation measures and, if possible, adopt devices with which to produce alternate energy that will serve to reduce one's own energy use or to produce for oneself. Any dollar not spent on energy is a dollar available to be spent and multiplied elsewhere in the local economy. Second, it is feasible to produce small-scale, technological devices locally that can reduce the use of energy.

What other products would offer opportunities? Clothing, shoes, and furniture are all areas that suggest themselves. How would they be chosen? We can imagine a very simple, straightforward process: Get together a small- to medium-sized group of friends and neighbors and ask them to bring their checkbooks for the last couple of years. (Assure them they will *not* be asked to write you a check.) Collectively, figure out what you spend money on. As typical consumers, what did (and do) you buy? Ask the question, what would we purchase that a group of talented, underemployed, local people might be able to produce locally? Make the workshop a community get-together and social event. Feed them something. Enjoy one another. We assure you, the ideas will flow.

It will certainly come up that certain products will not be produced locally, such as automobiles or electronics. This is true, and that is all right. The goal is not complete self-sufficiency in any case. Start with the easier, simpler products, and get the hang of it. Other ideas will come along. You cannot do everything at once anyway. However, we remind you that even the likes of General Electric, Microsoft, and Ford Motor Company started *somewhere*—often in someone's garage. You never know what undeveloped technological genius might be lurking out there in the intellectual-idea gene pool of your community. Give an opportunity to flourish. And that brings us to another issue, human requirements for resources.

Labor Force

From an economic viewpoint, requirements for the human engagement are in two categories. The first is for the general workforce; the second is for managerial and technical jobs. The first can be seen immediately as not a problem. Even if some technical training is required, and we hope it would be, the pool of talented people who are willing and able to work is exceptionally deep, and probably will remain so permanently. One recent job opening in state government immediately attracted 267 applicants, virtually all of them overtrained for the position. And the shame of it all was that the position could not even be filled due to last minute cuts in the budget. There are many who would welcome the opportunity to develop whatever skills might be necessary for a job earning a reasonable salary in a small-scale operation with a supportive work environment producing a product that everyone needs for a close and familiar market. Enough said.

The second set of skills involves the business and managerial talents and the technological expertise to set up and run a successful operation. There are successful businesspeople in every community. Top executives of major corporations like to foster the belief that they have a corner on the best managerial talents in the world. There are probably people in every community of any size who could run these corporations equally well, and with much more human compassion. Certainly, the best businesspeople in the community could be enlisted as consultants to share their experience and counsel, even if they are otherwise fully employed. Enthusiasm and support for the new enterprise would see to that.

For technological expertise, especially in setting up an operation, the community may have to look outside its boundaries. Remember, the whole effort seeks, in effect, to reverse some of the processes of globalization. There have been, by actual count, over 40,000 businesses closed and the jobs exported outside the country in the last decade in the United States. There would have to be unemployed or underemployed people tucked away in every corner of the nation who had, or have, the necessary knowledge of how such businesses function. They would not only relish but also need such an opportunity.

There is a technological challenge, however, and that is to produce efficiently on a small-enough scale so that the local population represents a market of sufficient size to remain profitable. This might require processes with more hand labor than normal, which ideally results in higher quality (biophysical principle 7 concerning trade-offs), and that is fine. The objective is to employ as many people as possible. The challenge for the industrial engineers is the opposite of what it might conventionally be: *Use as much labor as possible, and keep the scale small enough to avoid competition within the global economy.* Instead, enjoy your growing local monopoly.

Remember, the goal is to replace the use of products currently imported and purchased from global corporations. Although it may well be true that you cannot fight globalization, you can gradually go around it.

Financing the Vision

The question of financing the operation will quickly arise. Here, the bottom-line counsel to any community takes the same form as it does in most other areas: *You have more local resources than you think.* Most people are aware of the worldwide corporate power of the major product-oriented businesses. Companies like Exxon, General Electric, General Motors, IBM, Coca Cola, and a hundred others we could all name are widely viewed as purveyors of multinational corporate concentration—in effect, the stuff of which globalization is made.

But none of the business sectors implied in this list rival the financial sector for sheer monopoly-inducing concentration. The likes of Chase, Goldman Sachs, Citigroup, Merrill Lynch, Bank of America, and their few cohorts monopolize the world of finance much more tightly than any small group of the manufacturing firms in a particular sector could ever hope to accomplish for their range of products in the worldwide market.

And what has the financial sector done with all this power? Basically, it has ruined the rest of the economy for all of us, starting with insurance and housing, and spreading into virtually every economic sector. Without going into the eye-watering world of derivatives, hedge funds, foreclosures, credit default swaps, executive bonuses, and the greed that tied it all together, our conclusion is that the result has been perhaps the greatest upward shift in income the world has ever seen. Let us be clear, the main cause of the financial meltdown, which continues as we write, has been (and is) *inequality*. In other words, the financial sector has piloted a movement away from the necessary prerequisite of a sustainable world: *social equity*. Wall Street and its mode of operating is our greatest barrier to sustainability.

Where did they get all this money and power? We gave it to them. And we need to take it back. To be sure, the financial sector is ingenious in finding ways to expand capital values, but they need a great deal of it to start. That monetary wealth is extracted from every community in America. Most Americans have bank accounts, savings accounts, pension funds, mutual funds, or own a few shares of stock outright. Virtually all such assets, even though they were obtained through payroll deduction or some local friendly stockbroker, money manager, or financial planner—maybe a friend or relative, or in my case (RB) a former student—the money quickly finds its way to these financial powerhouse institutions.

Consequently, we might think, "Oh, we don't have the financial capability to start a manufacturing operation. We must look to outside financing." That is untrue. We simply need to get our own share back from Wall Street and use it. Savings are generated in every community. They have been expropriated, and we need to begin getting them back so that they can finance community health instead of community destruction. This is easier said than done, and we need improved mechanisms for doing that. State banks, community banks, and community credit unions are good examples of the types of institutions that would provide needed local finance.

What about the risk? Are we not safer going with big, sophisticated investors who have superior knowledge and who can diversify holdings and seek the best opportunities in the world? And how have they been doing for you lately? The safest economies of the world of the future are most likely to be local economies that take the largest steps down the path of community self-sufficiency and sustainability. The returns will not be spectacular (as the frenetic growth economy occasionally produces), but they will be secure and steady. True recession proofing means separating our fate, to whatever degree possible, from the global economy. Then we will be free to invest safely in ourselves.

Identifying the Market

The core of almost any business plan is its marketing plan. I (RB) have seen many business plans and have accompanied hopeful businesspeople as they engaged in conversations with bankers from whom financing was being requested. The response always comes down to this: "You've got a product here. How are you going to sell it? What's your marketing plan?" All the technical expertise and business acumen in the world are useless if you cannot sell the product.

This one is straightforward: *You are the market.* You and your friends and neighbors are already spending your income on food, energy, shoes, clothing, furniture, and a variety of other things. And all you are proposing to do is replace the alienating, disparity producing, environmentally destructive production mechanism for one that produces the products in a way that is community building, just, and stable. Simply the act of talking about it, imagining it, and setting it up creates the market. And once the market is there, it will never leave. It will only expand through the best and cheapest advertising method—word of mouth. And that is what I (RB) would tell the banker.

Finalizing the Vision

As a final note of encouragement, you are not alone. Models are out there in finance, economic initiatives, community organization, and a host of pertinent and related topics. The economic need is certainly everywhere. And fortunately, the core of commitment necessary to do something about it in an effective manner is present in virtually every town and city in the nation. We just have to free it, nurture it, and learn from one another. Mistakes will be made, and successes will be enjoyed. Communication, especially through the Internet, is quick and easy, which over time could allow us to avoid mistakes and quickly copy success. The first pickle out of the jar is the hardest to get.

In summary, there is the vision: Produce for yourself, using your own community resources. You are the investors, the workforce, and the market. The results will be a just and supportive community full of healthy relationships with people who are appreciative of the system in which they thrive. And it must all be supported by an economy that nurtures the necessities of life more than greed.

Although some suggestions for proceeding from this point may undoubtedly be gleaned from what we have already said, we conclude this work with our best counsel for taking the next steps in a personal journey toward the three pillars of sustainability inherent in the triple bottom-line framework—ecological integrity, social equity, and economic stability.

Endnote

1. The foregoing discussion is based on: Chris Maser. *Vision and Leadership in Sustainable Development*. Lewis, Boca Raton, FL. 1998.

11

Counsel for Getting There

The previous chapter sought to create a vision. Yet even if the vision is incomplete (as are all visions), it nonetheless spills over into recommended actions, which is the purpose of this concluding chapter. One more time: *Everything is connected to everything else.* And from our viewpoint, that certainly includes the various parts of this book and topic.

If, for instance, the vision is for a small-scale, locally owned business, the implied action is clear: *Go start one.* If the vision is for local finance, the action is also clear: *Close your account at Chase bank and open one at your local community bank or the antiglobalization credit union.* We are unsure, however, how much of our experience and counsel goes without saying and how much will need to be overtly stated.

If a vision melds into counsel, the leakage goes both ways: Counsel must be allowed to further refine the vision, just as vision oozes into counsel for action. We have done our best to focus this book on individuals and their internal worldview rather than on institutions or public policy, because people must change before institutions or policy will. As such, the most important action we can recommend is for you to *continue expanding your knowledge and vision based on that knowledge.* Our belief in the best of human nature leads us to conclude that if intelligent, other-centered (as opposed to self-centered) people begin to think *systemically,* they will make the kinds of decisions and initiate the kinds of actions that will lead toward social-environmental sustainability for all generations. We can only hope.

Resource Overexploitation

A major theme of this book has been that, as a culture and even as a species, we have overreached, or overexploited. Any final counsel we have to offer for personal action must rest squarely on that premise. In our drive to maximize the harvest of nature's bounty, we—especially in the United States—strive for an ever-increasing yield of products, and thus intensively alter more and more acres worldwide to that end. The Growth Ethic appears to demand this. The need, however, is a sustainable yield that cannot exist without first having a sustainable ecosystem, such as a forest or an ocean, to produce that yield.

Within the conventional approach, we tend to think it a tragic economic waste if nature's products, such as wood fiber or forage for livestock, are not somehow used by humans but are allowed instead to recycle in the ecosystem—to compost, as it were. And because of our paranoia over lost profits (defined as economic waste), we normally extract far more from every ecosystem than we replace. We will, for example, put capital in a crop but not in maintaining the health of the ecosystem that produces the crop. This type of expenditure is part of our Western, industrialized tradition and is thus ingrained in our culture.

According to a song popular some years ago, freedom is equated with having lost everything and thus having nothing left to lose. In a peculiar way, this sentiment speaks of an apparent human truth. When we are unconscious of a material value, we are free of its psychological grip. However, the instant we perceive a material value and anticipate possible material gain, we also perceive the psychological pain of potential loss. Historically, then, any newly identified resource is inevitably overexploited, often to the point of collapse or extinction. As we have often done in this work, let us proceed with two examples that lead to the formulation of some general principles.

EXAMPLE 11.1 PEARL OYSTERS

In the 16th century, the turkey-wing mussel replaced the pearl-oyster beds off the coast of Cubagua, Venezuela. The oyster's depletion was the result not only of overexploitation in a short period of time but also of the ecological stress the exploitation generated. Consequently, the turkey-wing mussel outcompeted the pearl oyster and thus prevented its recovery.[1] Depletion of the economically valuable species allowed the incursion of a competing species.

Such overexploitation, the specific details of which are unimportant, is an interaction of economics and environment. It has been experienced in many ways throughout history in numerous ecosystems with a multitude of products worldwide. The experience is inevitably based on the perceived entitlement of the exploiters to get their share before others do and to protect their economic outlay. The concept of a healthy capitalistic system is one that is ever-growing and ever-expanding, but in fact, such systems, and the ventures that fuel them, are no more sustainable biologically than the most fragile of the species that comprise the ecosystem. Consider a more modern example.

EXAMPLE 11.2 CHINESE TURTLE TRADE

The exploitation of turtles and tortoises for today's market in Asia contributes to a crisis in extinction of global proportions. Thus, it serves as a contemporary example of an unsustainable, capitalistic venture based on a biologically renewable resource thought to be inexhaustible.

Although mainland Southeast Asia has long been regarded as a mecca of diversity for turtles and tortoises, little is known about them in Laos, Cambodia, and Vietnam (formerly known as French Indochina), because there were no biological investigations prior to World War II. Since then, decades of civil unrest, political instability, and military conflict have largely prevented fieldwork. Nevertheless, turtles and tortoises face continuing exploitation for food and medicinal markets in Laos, Cambodia, and Vietnam, where hunters in rural villages capture them for local consumption or to sell to traders who periodically visit villages to purchase wildlife.

Although turtles are eaten locally and traded in Laos, Cambodia, and Vietnam, most are exported through Vietnam to markets in southern China to appease the extreme demand for the turtles' meat in soup and their shells for use in traditional Chinese medicine. Consequently, there are over one thousand large, commercial turtle farms in China that are collectively worth over a billion U.S. dollars.[2] The scale of these lucrative operations, especially those involving endangered species, poses a major threat to the survival of China's diverse turtle fauna.[3] This threat stems largely from the fact that turtle farmers are the primary purchasers of wild-caught turtles. They buy them to increase their overall stock of adult animals and to secure wild breeders. Wild breeders are important, because successive generations of farm-raised turtles exhibit a marked decrease in reproductive capability. The reliance on individuals captured in the wild demonstrates that turtle farming is not a sustainable practice. As populations of wild turtles decline, it will become increasingly difficult to supplement farm stock from the wild.

Even if turtle farming should crash, the depleted wild populations will still face overexploitation because there is an entrenched, cultural demand for wild-caught meat. The nutritional properties of wild animals are promulgated by the practitioners of traditional Chinese medicine and are thus deeply ingrained in the national psyche. Consequently, wild-caught turtles fetch significantly higher prices than those raised on farms, and no amount of captive breeding will decrease the insistence on obtaining wild turtles for consumption. Salmon and salmon farming in the United States present parallel situations to the Chinese turtle farms.

Historical Generalizations from These Examples

China is industrializing rapidly, and the escalation of turtle farming has followed the path of other capitalist ventures since the economic reforms of the 1980s. The fusion of China's growth with the utilitarian attitude of the Chinese toward nature clearly emphasizes the aforementioned fear of losing short-term profits, even at the cost of rendering long-term biodiversity unsustainable—as the history of economics has so often demonstrated. In the case of Chinese turtles, the farmers are grabbing the last vestiges of wild populations to process for the soup pot.

In effect, the tragic and often-repeated pattern is familiar. They are opting to earn profits as long as they can, regardless of the ecological outcome. In the long-term, therefore, the economic efforts serve a single function—to generate profit for a few entrepreneurs, despite the social-environmental consequences for all generations.[4]

As it is with any renewable natural resource, the nonsustainable exploitation has a built-in ratchet effect that works in this way. During periods of relative economic stability, the rate of harvest of a given renewable resource (e.g., wild turtles) tends to stabilize at a level that economic theory predicts can be sustained through some scale of time. Such levels, however, are almost always excessive because economists take existing unknown and unpredictable ecological variables and assume them to be known and predictable economic constant values in order to calculate the expected return on a given investment from a sustained harvest.

During good years in the market or in the availability of the resource, or both, additional capital outlay is encouraged in harvesting and processing, because this is the imperative of competitive economic growth and the root of capitalism. But when conditions return to normal or even below normal, the individual or industry, having overinvested, typically appeals to the government for help because substantial economic capital is at stake—including potential earnings. If the government responds positively, it encourages this cycle of continual overexploitation—the *ratchet effect*.[5] The ratchet effect is thus caused by unrestrained economic investment to increase short-term yields in good times and strong opposition to losing those yields in bad times.

Then, because there is no mechanism in our linear economic models of ever-increasing yields that allows for the uncertainties of ecological cycles and variability or for the inevitable decreases in yields during bad times, the long-term outcome is a heavily subsidized industry. Such an industry continually overharvests the resource on an artificially created, sustained-yield basis that is not biologically sustainable. When the notion of sustainability arises, the overexploiting parties marshal all scientific data favorable to their respective sides as "good" science and discount all unfavorable data as "bad" science, thereby politicizing the science and largely obfuscating its service to society. The result is that one generation steals from all future generations rather than face the loss of some idealized, potential income.

By generalizing from these examples, we extract six important lessons to be learned about the all-important issue of historical overexploitation of natural resources: (1) emphasize quality rather than quantity, (2) recognize that loss of sustainability occurs over time, (3) recognize that resource issues are complex and process driven, (4) accept the uncertainty of change, (5) stop perceiving loss as a threat to survival, and (6) favor biophysical effectiveness over economic efficiency.

In many ways, these are not new issues to this book. However, as points of summary, they deserve emphasis in that they offer a rock-solid foundation for any person to engage in their community and seek to promote effective change.

Lesson 1: Emphasize Quality Rather Than Quantity

Maximizing the quality of whatever we do with the Earth's finite resources will always conserve them, thereby spreading nature's wealth among more people and generations. Conversely, maximizing the quantity of any material withdrawn from the Earth's finite supply to feed the insatiable appetite of today's consumer economy can only squander nature's limited wealth. This said, we must choose unequivocally because we cannot maximize both quality and quantity simultaneously.

Lesson 2: Recognize That Loss of Sustainability Occurs over Time

A biologically sustainable use of any resource has never been achieved without first overexploiting it, despite the lengthy catalog of disastrous historical examples (such as the passenger pigeon and the North American bison or "buffalo") and the vast amount of contemporary data. If history is correct, resource problems are not environmental problems by nature, but rather human ones that we have created many times, in many places, under a wide variety of social, political, and economic systems.

Lesson 3: Recognize That Resource Issues Are Complex and Process Driven

The fundamental issues involving resources, the environment, and people are complex and process driven. The integrated knowledge of multiple disciplines is required to understand them. These underlying complexities of the biophysical systems *preclude a simplistic approach* to symptomatic, ecosystem manipulation. A straightforward regulation, which is a typical response, will almost never suffice, no matter how well-intentioned. In addition, the wide, natural variability and the compounding, cumulative influence of continual human activity mask the results of overexploitation until they are severe and largely irreparable within a human lifetime, if ever.

Our "management" of the world's resources is always to maximize the output of material products—to put conversion potential into operation. In so doing, we not only deplete the resource base and degrade habitat but also produce unmanageable, unintended outcomes often in the form of hazardous wastes. In unforeseen ways, these unintended products (euphemistically termed *by-products*) are altering the way our biosphere functions, usually in a negative way. Such unintended products include not only hazardous chemicals in our drinking water but also several veterinary drugs with which farmers inoculate their livestock that could kill scavengers—those species that clean our environment.[6]

Lesson 4: Accept the Uncertainty of Change

As long as the uncertainty of continual change is considered a condition to be avoided, nothing will be resolved. However, once the uncertainty of change is accepted as an inevitable, open-ended, creative process, most decision making is simply common sense. Consider that common sense dictates that one would favor actions having the greatest potential to be fixed when broken, as opposed to those with little or no potential. The important characteristic of ecosystem resiliency depends on the ability to repair a process, and this can be ascertained by monitoring results, followed by appropriate (common sense) modifying actions and policies.

Lesson 5: Stop Perceiving Loss as a Threat to Survival

We interpret the perceived loss of choice over our personal destinies as a threat to our survival. This sense of material loss usually translates into a lifelong fear of loss, which fans the flames of overexploitation through unbridled competition in the money chase and top-down, command-and-control management of natural resources.

As the human population grows, with a corresponding decline in the availability of natural resources, the pressure grows to increase top-down, command-and-control management of those resources. The fallacy of attempting to control ecosystems through management is that we humans are not in control to begin with—and never will be. We are, therefore, destined to fail whenever we attempt to enclose nature in a designer straightjacket: witness tornados, hurricanes, and forest fires.

Nevertheless, our socioeconomic institutions are inclined to respond to nature's erratic or surprising behavior by attempting to exert more direct control. Command and control, however, usually results in unforeseen consequences, both for ecosystems and for human welfare, in the form of collapsing resources, social and economic strife, and the continuing loss of biological diversity—along with the ecosystem services such diversity provides. Moreover, if the potential variability in an ecosystem's behavior is reduced through command-and-control management, as it often is in practice (such as damming and channeling river systems and replacing forests with monoculture tree farms), the system becomes less resilient than it was to perturbations, and the outcome can easily become an unwanted, biophysical disaster. Channeling the Mississippi River, for example, has resulted in more severe damage to human interests due to floods.[7]

People with the command-and-control ideology tend to think that any resource not converted into some sort of immediate profit is an economic waste, because that is the surest way to justify their existence to the growth-oriented general culture. They therefore view such activities as salvage logging and preemptive thinning as the only viable alternative to a biophysical disturbance, like a hurricane or a beetle infestation. Similarly, harvesting potential host trees in advance of insect infestations or disease or preemptively thinning or cutting forests in an attempt to control forest fires and thereby improve their resilience to potential stress and future disturbances may fill coffers in the short-term, but at the long-term, ecological expense of lost biological capital in the soil bank, among other things.[8]

Put simply, interactive systems perpetually organize themselves, with infinite novelty (biophysical principle 8), to a critical state in which a minor event can start a chain reaction that leads to destabilization and collapse, such as that of a forest following a fire. Following the disruption, the system will begin reorganizing toward the next critical state (e.g., forest succession), and so on indefinitely (biophysical principle 13).

Perhaps the most outstanding evidence that an ecosystem is subject to constant change and ultimate disruption, rather than existing in a static balance (biophysical principle 14), comes from studies of naturally occurring external factors that dislocate ecosystems. For a long time, ecologists failed to consider influences outside ecosystems. Their emphasis was on processes internal to an ecosystem even though what was occurring inside was driven by what was happening outside.

Climate appears to be foremost among these factors. By studying the record laid down in the sediments of oceans and lakes, scientists know that climate has fluctuated wildly over the last 2 million years, and the shape of ecosystems with it—witness what is going on today around the world. The fluctuations take place not only from eon to eon but also from year to year and season to season and at every scale in between; thus, the configuration of ecosystems is continually creating different landscapes in a particular area through geological time (biophysical principles 8 and 9).

Lesson 6: Favor Biophysical Effectiveness over Economic Efficiency

As an economy grows, natural capital, such as air, soil, water, timber, and marine fisheries, has inexorably been reallocated to human use via the marketplace, where economic efficiency rules. The conflict between economic growth and the conservation and maintenance of natural-resource systems is a clash between the economic ideals of *efficiency* and the realities of biophysical *effectiveness*, a distinction we have repeatedly stressed.

This economically driven divergence creates a conundrum, because traditional forms of active conservation require money, which, in the United States, is highly correlated with income and wealth. Therefore, the unfortunate question that arises in one form or another is as follows: Can we *afford* to protect the environment? That inappropriate question notwithstanding, the conservation and maintenance of biodiversity in all its forms will ultimately require the cessation of economic growth as perpetrated today. Perhaps, with wisdom, we could spend *less* rather than more in the strict economic sense, in ways that might increase both effectiveness as well as some appropriately defined version of efficiency.

Hereafter, the ultimate challenge will be first and foremost to maintain biodiversity, especially in the wake of globalization, because the number of threatened species is related to per capita, gross national product in five taxonomic groups in over 100 countries. Birds are the only taxonomic group in which numbers of threatened species decreased throughout industrialized countries as prosperity increased. Plants, invertebrates, amphibians, and reptiles show increasing numbers of threatened species with increasing prosperity. If these relationships hold, increasing numbers of species from several taxonomic groups are likely to be threatened with extinction as countries increase in material prosperity.[9]

If we choose to move toward a biologically adaptable landscape with the right attitude, any mistakes we make may become the wisdom of the future, because mistakes, which are simply misjudgments, stimulate the questions we ask and attempt to answer. But we must act while the Earth still has the strength and the resilience to survive in the face of ongoing errors, and while there still is an ecological "margin of safety" to allow a few more mistakes from which to learn. Remember, all we have to give the next generation is options. An option spent foolishly for short-term economic gains at the expense of long-term ecological sustainability is an option foreclosed forevermore.

Communities Must Actively Plan Their Own Futures

More than 3,000 years ago, sages who belong not just to India but to the whole world gave humanity one of the earliest spiritual treasures known to history, the *Rig Veda*. In it is a prayer addressed to all of us, a prayer that is the heart and soul of a shared vision—a means whereby people can plan their common future:

> Meet together, talk together.
> May your minds comprehend alike.
> Common be your action and achievement,
> Common be your thoughts and intentions,
> Common be the wishes of your heart,
> So there may be thorough union among you.[10]

A vision consists of the self-determination that people ideally want to move toward, not what they fear and want to move away from. As such, a vision, like the social-environmental sustainability of a community, is a perpetual work in progress. It is about doing the best we can to honorably push the limits of human possibility, a notion that is severely tested in times of turmoil. Pushing the limits of human possibility is a necessary condition of social evolution, because as Albert Einstein noted, "No problem can be solved from the same consciousness that created it."[11]

A shared vision of a sustainable future toward which a community can build creates confidence, agreement, and energy in equal parts. At a deeper level, it engages our imagination and helps to ferret out which questions need to be asked, how to word them, and when to ask them.

By engaging their imagination and sense of possibility of the ideal through initiatives, such as shared community vision, people who are concerned with social justice and the health of their environment can create an opportunity to confirm a more positive and sustainable future. Imagination, as Albert Einstein once noted, is more important than knowledge and is the most powerful tool for social change. However, community imagination and any subsequent actions, if they are to be healthy and productive, always proceed from a platform of values. Let's consider three categories of value: universal, cultural, and personal.

Universal Values

Universal (or archetypal) values reveal to us the human condition and inform us of our place therein. Through universal values, we connect our individual experiences with the rest of humanity (the collective unconscious) and the cosmos. Here, the barriers of time and place and of language and culture disappear in the ever-changing dance of life. Universal values must be experienced; they cannot be comprehended. Can you, for example, know a sunset? Fathom a drop of water? Translate a smile? Define love?

"Universal values are the timeless, unchanging desired of the human heart brought to different cultures at various times and ways throughout history. They remain ever at the center of human life, no matter where the hands of time are pointing—past, present, or future."[12] These are the truths of the human condition toward which all people aspire (such as joy, unity, love, and peace).

Cultural Values

Cultural (or ethnic) values are those of the day and are socially agreed upon. They are established to create and maintain social order in a particular time and place and can be highly volatile. Cultural values concern ethics and human notions of right and wrong, good or evil, in terms of customs, manners, and religion.

In culture, and in legal systems, we see reflected the ideas and behaviors that a society rewards or punishes according to their perceived alignment to its values. Hence, cultural values are for an individual a mixed bag, especially in a highly complex society that has lost its sense of family, community, and mythology. Today this is painfully apparent in the United States, where the capitalistic market and technological innovations are increasingly replacing the individual values of the simpler, more personally connected lifestyle of generations past.

Personal Values

Personal values are constituted by the private meanings we bestow on those concepts and experiences (such as marriage vows or spiritual teachings) that are important to us personally. These meanings are in large part a result of how we are raised by our families of origin and what of our parents' values we take with us in the form of personal temperament. These meanings may change, however, depending on our experiences in life and how much we are willing to grow toward psychological maturity as a result of our experiences. As such, individual values are reflected in such things as personal goals, humor, relationships, and commitments.[13]

Personal values, however they are derived, all have trade-offs (biophysical principle 7). Although it may not seem important at any particular moment in a given day, it is critical to know what values you personally want to safeguard for yourself and your children in the long run. This must be done in light of the recognized interplay with universal and cultural values. No act of community visioning can be successfully negotiated without this recognition, and many visions have failed because the participants ignored or manipulated these interactions, often to protect vested economic interests.

The amount of harmony and agreement or chaos and conflict in a community is a direct measure not only of how different people's perceptions really manifest but also of how committed they are to defending their individual points of view, regardless of their narrowness in scope. The purpose of a vision is to convert the present chaos into the greatest possible harmony for the collective benefit of the community as a whole through time, recognizing that each person's perception is part of the community's living culture. Living culture is thus embodied in the people, and each person is both the creator and the keeper of a unique piece of the cultural tapestry.

The important implication is that the freer we are as individuals to change our perceptions without social resistance in the form of ridicule or shame, the freer is a community (the collective of individual perceptions) to adapt to change in a healthy developmental or evolutionary way. On the other hand, the more people are ridiculed or shamed into accepting the politically correct ideas of others, the more prone a community is to the cracking of its moral foundation and to the crumbling of its social infrastructure.

Broad-Based Participation a Necessity

Our worldview is our way of seeing how the world works; it is our overall perspective from which we interpret the world and our place in it. There are, in the most general terms, two worldviews: the sacred and the commodity. One need not be religious in the conventional sense to hold a sacred view of life, because a sacred view focuses on the intrinsic value of all life. As such, it gives birth to feelings of duty, protection, and love, while emphasizing the values of joy, beauty, and caring, which in turn erects internal constraints to destructive human behavior against nature.

A commodity view of life is interested in domination, control, and profit and seeks to *gain the world* by subjugating it to the will of the industrial mentality through the economic system. At the core of the commodity worldview are several economic seeds, such as self-interest, the economy versus ecology dilemma, the growth/no growth tug-of-war, Rational Economic Man, and others. We have made the case that it is currently necessary, with respect to a commodity worldview, to protect the health of the environment in the present for the present and the future through external constraints placed on destructive human behavior that ignores, circumvents, and attempts to control how the biophysical principles function.

A vision does not create a single constraint, where there was none before. It cannot because everything in the world is already constrained by its relationship to everything else, which means that nothing is ever entirely free (biophysical principle 1). A vision determines the degree to which a particular socially chosen constraint is negotiable (biophysical principles 6 and 7). In addition, a vision forces a blurring of all interdisciplinary lines in its fulfillment, because the power of the vision rests with the people who created it and those who are inspired by it, not those whose sole job is to administer the bits and pieces of everyday life, as important as they might be.

What does *negotiable* mean here? It means to bargain for a different outcome, to cut the best possible deal. Can we, for example, negotiate with nature to give us sunnier, drier winters without flooding when we deem the winters too dark and wet? Can we negotiate for more rain during periods of drought? Well, we could try, but it would be to no avail. Nature does not negotiate; therefore, some of the conditions nature hands us are nonnegotiable (the biophysical principles). Thus, in some realms, we cannot cut a "better" deal, one more to our liking.

Because a community's visioning process is public, it gives the people the right to comment on all aspects of the process (from creating the vision through its implementation and monitoring), which, in effect, places control of the process directly in the hands of the people, should they choose to accept responsibility for the outcome. The responsibility for the outcome demands an understanding of and exacts the accountability for how people accept the social constraints dictated by the vision and their compliance with the limitations of nature's biophysical principles.[14]

Need for Bottom-Up Thinking

As we have said before, there is great power in learning to reframe negatives into positives. In doing so, the participants in creating a shared vision not only understand their community from several vantage points but also understand that much of the confusion in communication comes from trying to move away from negatives. Trying to move away from a negative precludes people from saying what they really mean because they are focused on what they *do not want*. As long as people express what they do not want, it is virtually impossible to figure out *what they do want*.

To facilitate community visioning, therefore, let us construct an example that involves a principle that is tacitly present in much that we have said, but which needs to be reinforced: *Bottom-up solutions are always preferable to top-down control*. Moreover, the cumulative effect of like-minded, little ideas is to culminate in a big (systemic) idea, and *big ideas are sometimes necessary to promote big change*. In an industrial, action-oriented culture such as ours, many authors have insightful and erudite analyses of a problem of one type or another. Certainly, there are many books and articles on the problems of the economy or the environment. But, feeling an unspoken challenge on the part of the reader to present viable, systemic solutions, the author falls into a common trap in response to the perceived: "I see the problem. What can we do about it?" Then, more often out of habitual, symptomatic thinking than anything else, the authors almost invariably produce a litany of top-down, symptomatic measures, such as the following:

- Increase federal infrastructure spending
- Sign the Global Warming treaty
- Extend unemployment benefits
- Pass cap-and-trade legislation
- Fund national, high-speed rail
- Ban offshore drilling

Do not get us wrong. These may or may not be good ideas. Nevertheless, such recommendations (or their opposite), however well argued and defended, encourage people to proceed within the conventional mindset. The approach does not facilitate the kind of thinking outside the box that we have attempted to promote with this work. In fact, we suspect that many excellent works, based on outstanding research, simply join the existing top-down debates on one side or another of important issues, and ultimately serve merely to harden existing opinions and do nothing to diminish existing political and cultural conflict, as the following example elaborates.

Example: The Minimum Wage

Suppose we observe that a great many low- and moderate-income people are working very hard in our system for extremely low wages and are thus barely making ends meet, if that. They can sometimes scarcely pay the rent or put food on the table for their kids. As a consequence, they cannot afford much when it comes to entertainment, cultural activities, or even the requirements of basic transportation: the single mother who works at McDonald's, the recent immigrant doing laundry in a nursing home, or the night watchman at an industrial warehouse. (Let us assume that the example has not yet wrested you out of the realm of reality.)

Further, we note that many of these people in such backwaters of the service sector are earning the minimum wage. And this does not even consider the human suffering of more than 10 million people in the United States who are currently unemployed and receiving unemployment compensation, which amounts to less than the minimum wage.

What should we, as concerned citizens, do about it? Well, the first observation might be that the minimum wage is too low, and the conclusion is that it must be raised. At this point, we cannot only critique the potential effectiveness of a top-down approach as compared with a bottom-up approach to the perceived problem but also of a typical conservative or liberal mindset.

Top-Down Approach and the Minimum Wage

In a top-down approach, it is expected that the federal government will act, and the first response would be that the minimum wage should be raised. Most of us do not have the power or prestige to write legislation or go to Washington and testify (if a bill to increase the minimum wage could even get introduced). Thus, the upper limits of activism on the issue probably stop at writing or calling your congressional representative, or your U.S. senator to indicate your support for increasing the minimum wage.

Do not misunderstand us; calling your congressional representatives is always a good idea, because they need to hear more often from their constituents as it is. But, as an effective step in solving the problem of the working mother, it has little hope of being effective. If there is no current bill pending, the contact will fall on deaf ears. There is another problem, the actions indicated above are likely to be the response from a progressive or liberal-minded person.

The conservative response to wage insufficiency in the lowest-paying sectors of the economy would not be to address the minimum wage in the first place, because that is seen as a distorting regulation. Instead, the likely response would be to do away with such regulations entirely (or even *lower* the minimum wage) and cut taxes in order to stimulate business and create new jobs.

If such a bill to raise the minimum wage is introduced, the conservative business interests would therefore oppose it, while progressives favor it, and an important point for our analysis here is that the top-down approach not only accomplishes little but also creates additional political and cultural conflict.

As an economist, I (RB) need to add another perspective to this particular issue. Raising the minimum wage is a favorite example of economists of the workings of supply and demand. Any downward-sloping demand curve says that if the price of something is raised, less of it will be consumed. Thus, conventional economic wisdom says that if the wage rate is (artificially) increased, unemployment will result. People who would have been employed at the lower rate will be let go at the higher wage level. Businesses either would not be able to afford it or would choose not to. At this point, the traditional economist draws a few lines on the graph and says, "This much unemployment will occur," and then moves on to the next topic.

At times during my career I have spent hours over lunch asking my colleagues why they suppose data never show that an increase in the minimum wage causes unemployment. Over the years, as the minimum wage has many times been raised, it has never been statistically measured to have led to any unemployment whatsoever. I will not go into the answers I have received, because there were no real answers. They simply continued believing what they believed. But, I have some suggestions.

First, it is very likely that the business in question really *was not* paying the workers all it could afford, because the business had all the power in the model anyway, and was, therefore, simply paying as little as the business could legally get away with. So, after the minimum wage is increased, the business owners need not fire any of the employees because they cannot afford them. Second, when wage rates are increased, for whatever reason, more purchasing power is injected into the economy, especially on the part of people who have to spend virtually all of their income immediately. There will be more demand for goods and services, and thus more demand for the labor to produce them. The minimum-wage increase may be part of the reason for a *higher* demand for labor.

The failure of economics and economists to acknowledge this possibility is a major shortcoming and exemplifies a criticism we emphasized earlier. It fails to acknowledge the whole system, that there really is no independent or dependent variable, and thus assumes things are constant that do not and cannot remain constant (biophysical principle 1). When supply/demand analysis is employed, for instance, it is common to add the Latin phrase, *ceteris paribus*, or "all other things equal." This is a convenient assumption for the methodology of the moment, but unfortunately it does not represent the world as it is (again, biophysical principle 1: *everything is a relationship*).

Before examining the *bottom-up* approach, two other points are in order. First, in our legislative example, the discipline of economics, intentionally or not, would come down squarely on the side of conservatism. The business interests, eager to hold down labor costs, would aggressively invoke the support of "hard-headed economic logic" in opposing a minimum-wage hike. This is one of the best examples of how economics, one more time, ends up a reactionary discipline, even when it is being used incorrectly by some vested interest as it suits their needs.

Second, what can we say, simply on humanitarian grounds, about a methodology that would support denying a struggling worker 25 or 50 cents more per hour? The prospect of a lawyer/lobbyist, who makes $200 or $300 per hour testifying before a congressional committee for their corporate client to contend that their multinational boss cannot afford to pay a single mom another portion of a dollar per hour is ludicrous. Rational Economic Man is alive and well, but he has no business holding elected office.

Bottom-Up Approach and the Minimum Wage

Returning to the plight of our struggling worker, what action is appropriate from the bottom-up viewpoint? Building on our Chapter 10 vision of a local economy based on the necessities of life, one would be wise in seeking to be part of a movement to create local businesses producing for their own community. Then, instead of a future of cookie-cutter, minimum-wage jobs for an international company or fast-food chain, the community could begin working toward a range of better-paying and meaningful jobs creating the goods that you, your family, and neighbors all want and need.

Admittedly, this sounds utopian and slow and seems to do nothing for the national minimum-wage problem. However, it is entirely consistent with the vision, and the challenge to become active in your community. Further, if one community manages to create a successful operation, it *will* (not just *may* in these trying economic times, but *will*) quickly prove a replicable model for many other communities. And this will put upward pressure on the national minimum wage in the healthiest possible manner. If such a movement should sweep the nation, with a variety of different, widely used products, we can envision a day when a "minimum wage," as we know it, would cease to be a necessity. We can only hope.

Finally, such an action plan, albeit ambitious, would be looked upon favorably by progressives who wish to wrest control away from large corporations, and who have deep concern for the plight of common people. Conservatives who promote jobs, local control, and individual initiative would also greet it with the utmost enthusiasm. Both perspectives would favor the decreased need for implied regulation, as well as many other of the stated advantages as indicated by our vision. In short, bottom-up cooperation leads to the type of political harmony that top-down control disrupts. An economic arrangement would finally exist—for the first time in over two centuries—that once again promotes ecological integrity, social equity, and economic stability as opposed to alienation, inequality, and environmental degradation.

A Final Word on Growth

At almost every stage of this book, the analysis has run up against the *Growth Ethic,* in one way or another. We conclude with some brief, final comments. Much of what we say here has been said elsewhere, but we organize it differently. The question could arise, especially from the skeptical reader: If growth is so terrible, why and how has it managed to become our virtual public religion? Has modern society been on a collective disastrous course for over 200 years now, or are there advantages to economic growth that you are not acknowledging?

Fair questions, and remember, the topic of this final chapter is our best (*personal,* not *policy*) counsel for proceeding into the world, where social-environmental-economic sustainability might become the prevailing worldview. All authors are biased, and all statements of any merit are value based, so once more we reiterate our overriding bias: If you do not live in a world where *sustainability* is the pervasive order of the day in all that you do, you live in a failing world. And it is our overriding purpose to do all we can to promote social-environmental-economic sustainability for our communities, our world, ourselves, our children and grandchildren, and all those who follow.

Growth affects us in many ways. Accordingly, we summarize by briefly discussing a few of the pros and cons of growth, as we see them, in five different and important categories: community, national economy, politics and culture, environment, and personal.

Community

Pro: Growth is held to be a dynamic factor that keeps a community from stagnating and, given the obvious mobility of Americans, accommodates the inevitable influx of new people, both by birth and in-migration. It provides new economic opportunities that allow children to remain in their communities as they grow up. It creates higher incomes and salaries, which fund, privately and through taxes, any amenity the community might want. It creates jobs directly, through the building trades.

Con: The changes that supposedly avoid stagnation often result in alienation, whereby features and advantages that people historically perceive are destroyed by growth. Growth is perceived to serve the interests of a certain portion of a community only, and thus becomes polarizing and destructive to civic cooperation. The new incomes and new taxes rarely fund the changes or the amelioration of the problems caused, and further controversy results, including inadequate public-sector budgets. Building and expansion exacerbate the consumer culture, in addition to palpable land-use impacts that degrade the quality of life, which causes long-time residents have less interest in volunteering and contributing to the social capital of the community.

National Economy

Pro: A growing economy generates more jobs and thus higher ability of individuals and families to support themselves, which increases individual happiness and well-being. Tax collections will be higher also, thus allowing the public sector to more easily pay for whatever people want and still be perceived as adequate. Our international strength and ability to defend ourselves in a dangerous world are supposedly higher with a dynamically growing economy. Broad-based prosperity is depicted as the result.

Con: To a point, some of these arguments may have some validity, but it is never enough. The globalized-growth economy, augmented by incessant technological change, seems never to create jobs where one wants them, and never with high enough incomes. Families are stressed when both the private and public sectors fail them not only in bad times but also in good times by forcing both adults to work to make their financial ends meet, thus fragmenting the family. Efforts around the world to command resources needed to keep growing are simultaneously destroying ecosystems and actually undermining international credibility, which makes the world a more dangerous place. The most pervasive result, as we repeatedly stressed, is the disastrous inequality that growth inevitably leads to, and which our macroeconomics inadequately recognizes. This causes the unfortunate "have/have not" dichotomy at all levels, from community to global, and undermines the cooperation so badly needed in an overcrowded world.

Politics and Culture

Pro: The political system is assumed to be working well only if the economy is healthy, which is interpreted as growing. For any political campaign in an unhealthy economy, the economy is assumed to be the dominant, if not the only, issue, as evidenced by Bill Clinton's famous admonition to himself as a campaigner: "It's the economy, stupid." Growth is assumed to help everyone. Consequently, an expanding economy is assumed to serve the best interests of the middle-class and democracy. Moreover, if the economy is robust, incumbents are accorded an easier reelection but are in trouble during bad times.

Con: Clearly, no incumbent politician is totally responsible for either a growing or the stagnant economy, despite the unfortunate superficiality that inference implies for our election processes. Environment and a host of other important issues are ignored because of the obsession with economics. Materialism and greed allow taxes to be effectively billed as a waste of our private purchasing power rather than the wherewithal to purchase vital necessities. The thoughtful citizen is turned off, and many others become myopic, single-issue voters. Better public education and universal health coverage take a back seat and presumably await an improved economy so that we can afford these basic human necessities—even though, like the Growth Ethic, we never seem to get there.

In short, pervasiveness of the Growth Ethic in our politics leads to misplaced priorities, superficiality, lack of participation and interest in the act of choosing our elected leaders, and heightened cultural conflict as inequality worsens. Campaigns become strident battles marked by meaningless sound bites, instead of thoughtful arenas for periodic public debate on cultural priorities. This said, the universally recognized effects of money in politics are a direct result of the emphasis on materialism and continual growth, and dramatically threaten nature's ecological services on which we all depend and the unencumbered democratic equity they allow.

Environment

Pro: Any arguments that linear growth is good for the environment require a strictly "mitigation" frame of mind. In other words, the fact that ecosystems become degraded under economic competition requires that we mend what is broken, nurture the processes that serve us, and maintain their productive capacity for future generations. Of course, this obligation requires a certain amount of monetary capital. Little can be said about *advantages* for the environment under a mandate of continual economic growth, because such a mandate is based on the tacit assumption that we cannot really harm a natural system and thus will never run out of some needed resource. Therefore, economic growth conveniently ignores any and all environmental consequences. Nevertheless, the fact is that we have no choice but to alter the environment in securing our living simply because we exist and use energy. And we have a right to do so. But, how and why we alter the environment, the level of consciousness we display in how we treat nature, is altogether a different issue—and is a matter of choice.

Con: As this book implies, environment is the area in which we practice the most strident informed denial. Any argument that a healthy economy is needed for a healthy environment must single-mindedly assume that money is the answer to everything. Human health matters little. The quality and type of energy used is unimportant. Community cohesion makes no difference. And, most importantly, the environment is seen as an external thing to be taken for granted or managed as opposed to our vital source of support that is to be nurtured, both economically and psychologically. It is in this arena that our overriding message must be delivered: *Permanent, linear, economic growth is biophysically impossible.*

Personal

Pro: Economic growth supposedly creates opportunities for careers, higher income, and the *good life*. Culture, the arts, and personal enrichment are acknowledged as important, but economic success is assumed to be a necessary precondition. Although one can act altruistically or serve the community, philanthropy requires that economic wherewithal must come first.

Con: On the one hand, great philanthropists and foundations provide compelling examples that superficially support these arguments. On the other hand, many of the fortunes that could be referenced in this regard were built on the backs of the poor, and often accompanied—or even caused—a great deal of the inequality and class conflict. Their foundations now succumb to the paradox of purporting to treat the misery they initially caused. The main problem that most of us can relate to, however, is the mindset.

If we accept the mandate that more is better and continual growth is essential in forming our personal philosophy of life, then we mount a treadmill from which we can never escape. If the premise of our superficial, materialistic culture is that you always need more to be successful, enough will never come. Acquisitiveness supercedes cultural and artistic expression. A workaholic approach can dominate rest, relaxation, and building personal relationships. Cultural icons tend to be wealthy people, and getting ahead becomes an obsession. But, when an economy dominated by such personal agendas fails to perform, the personal sense of bitterness over the lack of purchasing power and the accompanying loss of control accentuate the palpable inequality. This combination of factors not only serves up a recipe for collective cultural and political conflict but also imposes damaging effects on the individual psyche from the assumed message of social failure.

Summing Up

This has been a very quick and incomplete listing of conditions in a variety of areas, as they exist in a world dominated by the Growth Ethic—all evaluated from the platform of a worldview emphasizing social-environmental-economic sustainability. Remember, these conditions are viewed from our vantage point with the goal of attempting to avert a growing and irreparable crisis we see looming into the future for all generations. You must, however, make up your own mind and choose your own path. You must go into the world and create the future that you want; if your actions are the product of a growing knowledge of cause and effect based on the biophysical principles outlined in this book, you have more power than you realize to shape a positive contribution for your having been here.

Whatever set of beliefs and paths you choose, always remember the legacy that we all owe to future generations. This is to say, you must understand the environmental and economic circumstances to which you are committing the future, because if your choices create a deficit in terms of either a child's future options or the ecosystem's productive capacity, it is analogous to "taxation without representation," and that goes against everything a true democracy stands for. The choice is yours. How will you choose?

Endnotes

1. Aldemaro Romero. Death and taxes: the case of the depletion of pearl oyster beds in sixteenth-century Venezuela. *Conservation Biology* 17 (2003):1013–1023.
2. Shi & Provincial Forestry Bureau for Endangered Species, Import and Export Management Office of China, July 2007, unpublished data.
3. Norman Myers. The extinction spasm impending: synergisms at work. *Conservation Biology* 1 (1987):14–21.
4. The preceding discussion of the trade in turtles and tortoises is based on: M. D. Jenkins. *Tortoises and Freshwater Turtles: The Trade in Southeast Asia.* TRAFFIC International, Cambridge, UK. 1995; Le Dien Duc and S. Broad. *Investigations into Tortoise and Freshwater Turtle Trade in Vietnam.* Gland, Switzerland, and Cambridge, UK; Species Survival Commission, International Union for the Conservation of Nature, 1995; M. Lau, B. Chan, P. Crow, and G. Ades. Trade and Conservation of Turtles and Tortoises in the Hong Kong Special Administrative Region, People's Republic of China. In: *Asian Turtle Trade: Proceedings of a Workshop on Conservation and Trade of Freshwater Turtles and Tortoises in Asia*, P. P. Van Dijk, B. L. Stuart, and A.G.J. Rhodin (eds.), Chelonian Research Monographs, no. 2, 39–44 (Lunenburg, MA: Chelonian Research Foundation, 2000); P. P. Van Dijk, B. L. Stuart, and A. G. J. Rhodin (eds.). *Asian Turtle Trade: Proceedings of a Workshop on Conservation and Trade of Freshwater Turtles and Tortoises in Asia.* Chelonian

Research Monographs, no. 2 (Lunenburg, MA: Chelonian Research Foundation, 2000); H. Shi and J. F. Parham. Preliminary observations of a large turtle farm in Hainan Province, People's Republic of China. *Turtle and Tortoise Newsletter* 3 (2001):2–4; R. H. P. Holloway. Domestic trade of tortoises and freshwater turtles in Cambodia. Linnaeus Fund Research Report. *Chelonian Conservation and Biology* 4 (2003):733–734; H. Shi, Z. Fan, F. Yin, and Z. Yuan. New data on the trade and captive breeding of turtles in Guangxi Province, South China. *Asiatic Herpetological Research* 10 (2004):126–128; Haitao Shi, James F. Parham, Michael Lau, and Chen Tien-His. Farming endangered turtles to extinction in China. *Conservation Biology* 21 (2007):5–6.

5. The preceding two paragraphs are based on: Donald Ludwig, Ray Hilborn, and Carl Walters. Uncertainty, resource exploitation, and conservation: lesson from history. *Science* 260 (1993):17, 36.
6. Stuart W. Krasner, Howard S. Weinberg, Susan D. Richardson, and others. Occurrence of a new generation of disinfection byproducts. *Environmental Science & Technology* 40 (2006):7175–7185; Susan Milius. Birds beware. *Science News* 170 (2006):309–310.
7. The preceding three paragraphs are based on: C. S. Holling and Gary K. Meffe. Command and control and the pathology of natural resource management. *Conservation Biology* 10 (1996):328–337.
8. The preceding two paragraphs are based on: David R. Foster and David A. Orwig. Preemptive and salvage harvesting of New England forests: when doing nothing is a viable alternative. *Conservation Biology* 20 (2006):959–970.
9. Robin Naidoo and Wiktor L. Adamowicz. Effects of economic prosperity on numbers of threatened species. *Conservation Biology* 15 (2001):1021–1029; Oliver R. W. Pergams, Brian Czech, J. Christopher Haney, and Dennis Nyberg. Linkage of conservation activity to trends in the U.S. economy. *Conservation Biology* 18 (2004):1617–1623.
10. Translated from the original Sanskrit by H. H. Wilson. Edited by W. F. Webster. *Rig-Veda-Sanhitá: A Collection of Ancient Hindu Hymns, Constituting Part of the Seventh and the Eighth Ashtaka*. Trübner & Co., London. 1888. p. 415.
11. Albert Einstein. http://coolquotescollection.com/1403/no-problem-can-be-solved-from-the-same-consciousness-that-created-it-we-must (accessed on December 13, 2010).
12. Laurence G. Boldt. *Zen and the Art of Making a Living*. Penguin/Arkana, New York. 1993.
13. The preceding discussion is based in part on: Laurence G. Boldt. *Zen and the Art of Making a Living*. Penguin/Arkana, New York. 1993.
14. The preceding discussion of a vision is based on: Chris Maser. *Vision and Leadership in Sustainable Development*. Lewis, Boca Raton, FL. 1998.

Appendix

Common and Scientific Names of Plants and Animals Mentioned	
Mollusks	
Pearl oyster	*Pinctada imbricata*
Turkey-wing mussel	*Arca zebra*
Reptiles	
Tortoises	Testudinidae
Turtles	Chelonia
Birds	
Passenger pigeon	*Ectopistes migratorius*
Mammals	
North American bison	*Bison bison*

Index

E

Q

R

Y

Z